Advanced Topics in Science and Technology in China

Volume 64

Zhejiang University is one of the leading universities in China. In Advanced Topics in Science and Technology in China, Zhejiang University Press and Springer jointly publish monographs by Chinese scholars and professors, as well as invited authors and editors from abroad who are outstanding experts and scholars in their fields. This series will be of interest to researchers, lecturers, and graduate students alike.

Advanced Topics in Science and Technology in China aims to present the latest and most cutting-edge theories, techniques, and methodologies in various research areas in China. It covers all disciplines in the fields of natural science and technology, including but not limited to, computer science, materials science, the life sciences, engineering, environmental sciences, mathematics, and physics.

This book series is indexed by the SCOPUS database.

If you are interested in publishing your book in the series, please contact Dr. Mengchu Huang (Email: mengchu.huang@springer.com).

More information about this series at http://www.springer.com/series/7887

Yun Zhang

Place Making in International Practice of Landscape Architecture

A Study of Australian Practices in China

ZHEJIANG UNIVERSITY PRESS
浙江大学出版社

Springer

Yun Zhang
Zhejiang University
Hangzhou, Zhejiang, China

ISSN 1995-6819 ISSN 1995-6827 (electronic)
Advanced Topics in Science and Technology in China
ISBN 978-981-16-2441-4 ISBN 978-981-16-2442-1 (eBook)
https://doi.org/10.1007/978-981-16-2442-1

Jointly published with Zhejiang University Press, China
The print edition is not for sale in China Mainland. Customers from China Mainland please order the print book from: Zhejiang University Press.

This Springer imprint is published by the registered company Springer Nature Singapore Pte Ltd.
The registered company address is: 152 Beach Road, #21-01/04 Gateway East, Singapore 189721, Singapore

Contents

Abbreviations

AILA	Australian Institute of Landscape Architects
APSFADTRC	Australian Parliamentary Senate Foreign Affairs, Defence and Trade References Committee
Austrade	The Australian Trade Commission
CHSLA	The Chinese Society of Landscape Architecture
CLA	Chinese Landscape Architecture
EBOCWJ	Editorial Board of China Wuxi Journal
EIA	Environment Impact Assessment
IFLA	International Federation of Landscape Architects
RDUDP	Ron Dennis Urban Design Planning and Business Consultancy
WMCPC	Wuxi Municipal Committee of Planning and Construction of the Li Lake Area
WAA	Williams, Asselin, Ackaoui & Associates
WPB	Wuxi Planning Bureau

List of Figures

List of Tables

Chapter 1
Introduction

Contacts, chains of command, personal interlinkages, and relations of social power and domination are increasingly stretched out around the surface of the planet. And in the midst of this global connectedness, place and culture are restructured.
Massey and Jess (1995: 1)

1.1 Background

The topic of this research is international practice in landscape architecture. In the contemporary era, landscape architectural practice exists in an international market of planning and design services. In this sense, the profession seems ubiquitous. However, the extent and specific quality of its contributions are not adequately understood. History shows that international practice in landscape architecture has occurred for many hundreds of years. Garden design was practiced by talented amateurs, such as European merchants and missionaries in China in the seventeenth and eighteenth centuries. One prominent example was the Chinese imperial garden: Long Spring (*Chang-chun*) Garden (Fig. 1.1), designed by the Italian Jesuit Giuseppe Castiglione (1688–1766), who acted as a court painter and lived in China for fifty years. These sporadic exchanges across cultural boundaries are interesting in themselves but have left an eclectic trail of garden design artefacts.

History also shows that international practice in landscape architecture was conducted by experts in landscape gardening. In the late nineteenth and early twentieth centuries, a peak time for Western colonisation, parks and buildings were designed and built in the territories leased by western colonisers, who were also investors, such as those in Hong Kong and Shanghai. During this period, the western term 'Park' and its modern recreational meaning were introduced in China. One prominent example is the early parks along the Bund in Shanghai, which had been originally designed for families of western colonisers and were constructed in 1866–67 (Fig. 1.2). These parks had been designed and maintained by the British landscape gardener George R. Corner for nearly eight years after their construction in 1867.

Y. Zhang, *Place Making in International Practice of Landscape Architecture*,
Advanced Topics in Science and Technology in China 64,
https://doi.org/10.1007/978-981-16-2442-1_1

Fig. 1.1 Castiglione's design for the Chinese imperial garden in the 1700s. *Source* A Pictorial Record of Chang-chun Garden (Museum of Dong-shan-sheng 1931)

Fig. 1.2 The public parks along the Bund in Shanghai, China, 1876. *Source Study on Formation Mechanism and Characteristics of Early Parks and Open Spaces along the Bund* (Wang 2008: 94)

During that period, the use of greenhouses, styles of English Victorian gardens, emphasis on plants' horticultural values, and species from Australia and England were introduced (Wang 2008). However, the extent to which such styles transformed or influenced in any way the design of public spaces on a broad scale is questionable. They are instead historic artefacts of the time and interesting to individual garden experts.

In the late twentieth century, international practice in landscape architecture seems distinctly different. For a start, it is less concerned with amateurs or stand-alone designers and more concerned with a suite of professional consultancy services. Moreover, the process of providing services in the present day is often in a relatively short time, being only months or years, rather than decades. The noticeable impact of such organised professional activities, however, seems far broader than the practice of travelling amateurs and individual garden experts. Further, the shift from colonial times suggests that contemporary international design practice might be in demand for mutual benefits and exchange, rather than the one-way imposition of forms and functions as manifested during colonial times. Such historical differences appear to be related to the contemporary tide towards global connectedness, as indicated by Massey and Jess (1995), which has the potential for restructuring place and culture.

In recent decades, international practice in landscape architecture has been responsible for planning and designing a diverse set of projects, from everyday places to recreational waterfronts and monumental landscapes, from land-use planning to the design of public parks and open spaces. Reports of these practices are widely available from the websites of firms providing landscape architectural services internationally; from professional journals in countries not only supplying but also requesting such services; as well as from newspapers, broadcasts, TV shows and the internet. For example, the core of the 2000 Sydney Olympic site in Australia was master planned by the US-based firm Hargreaves Associates; in China, the 2008 Beijing Olympic site was master planned by the US-based firm Sasaki and Associates; and in the United Kingdom, the forthcoming 2012 Olympic site in London was master planned by the US-based firm EDAW. Beyond Olympic sites, international practice in landscape architecture has been engaged in the planning and design of new cities, waterfronts, urban parks and the design of residential landscapes. Using Australia as an example, its destination of international landscape architectural services has extended from Southeast Asia to East and West Asia, such as Malaysia, China and Dubai.

In China, foreign landscape architectural practices have been sourced from across the world, including America, Australia, Canada, Europe, Japan and Singapore. Among the foreign landscape architectural services that have been provided in China, many are from Australia. For example, in the 2000s, the firm Hassell was involved in the planning of Ningbo New Town in Ning-bo, and Singapore Hi-Tech Park in Nanjing. Also in the 2000s, the firm Place was involved in the design of residential landscapes of Nine-Brook Rose Garden (*Jiu-xi-mei-gui Yuan*) in Hangzhou in South-eastern China and Qingdao Coastal walkway in North-eastern China. More recently, the firm Aspect was involved in the planning and design of parks and eco-resorts in several southern and northern cities in China in 2004 and 2005 (Fig. 1.3). These Australia-based firms include those that specialise in landscape architecture

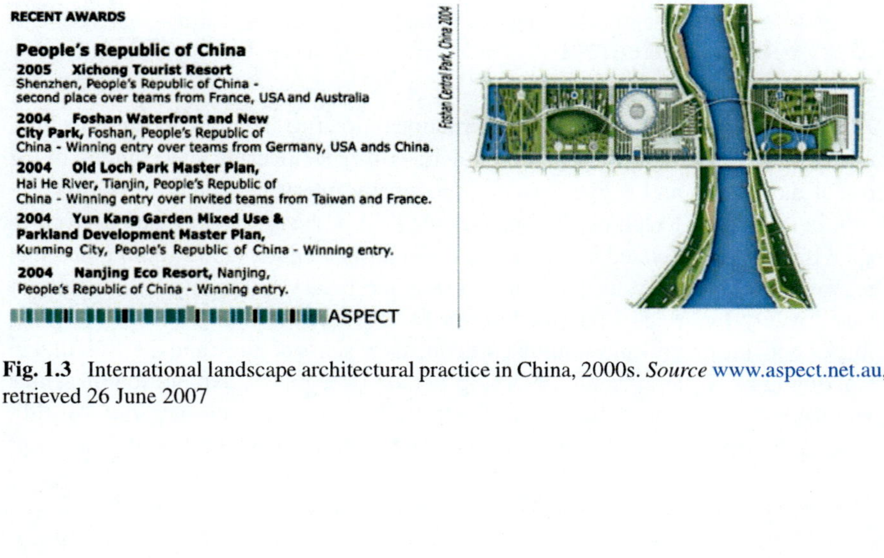

Fig. 1.3 International landscape architectural practice in China, 2000s. *Source* www.aspect.net.au, retrieved 26 June 2007

and some that are multidisciplinary and development oriented, but all claim to be experienced in providing the full suite of landscape architectural services.

These brief examples suggest a number of trends. First, international landscape architectural practice has increased in recent decades, expanding its working territory offshore. Second, international landscape architectural practice often appears to be associated with other planning and design professions. Third, countries providing services internationally appear to have undergone the institutionalisation of landscape architecture as a modern profession. Among these trends, however, the spread of landscape architectural practice internationally does not show a distinctive pattern, but seems highly specific to particular countries and their social or cultural circumstance. In some instances, international landscape architectural practice has spread between relatively similar cultures, such as American practices in Australia and Singapore or Japanese practices in China. Other examples suggest it spreads between remarkably different cultures, such as American practices in Malaysia and Australian practices in China. However, not many examples show the spread of international landscape architectural practice from developing countries to the developed.

As part of such phenomena, the extent and specific quality of international landscape architectural practice seems to be not adequately understood. First, how is the practice of landscape architecture positioned in the international market of planning and design services? When clients and practitioners both regard certain international planning and design services as landscape architecture, what are the roles of landscape architects? What are the types of projects and types of services provided? Second, when landscape architects are positioned as international practitioners, how do they understand a place and deal with its characteristics? When landscape architecture's approaches to place and practice suggest a primary goal to sustain the characteristics of place, these approaches seem to be challenged by cultural difference and geographical distance manifested in international practice. The way in which the characteristics of place are understood and sustained in such a context is not sufficiently understood.

1.2 Problem Statement

A theoretical gap exists concerning international practice in landscape architecture. On the one hand, theories provide guidelines for deciphering the complexity of present day places and suggest ways to positively intervene in their re-formation. Early in the 1970s, Relph (1976) and Tuan (1977) both argued for place as lived-in experience with associated rich meanings. Similarly, Seddon (1972: 262) noted that the way the community feels about and uses the landscape underpins its sense of place. Twenty-five years later, Seddon (1997: 136) further highlighted the inseparability of both the natural and cultural components of sense of place and the difficulty in resolving these. In architecture, Norberg-Schulz (1980: 201) exploration of genius loci also sees sites and buildings as'existentially rooted in everyday life[-]world', stressing the spiritual component of place and arguing against the proliferation of international styles and modern technology. These theories all suggest that impersonal and efficiency-oriented approach in planning and design are often responsible for the loss of places.

In recent decades, the characteristics of place are frequently investigated in relation to globalisation. Using Harvey's (1990: 350) words, the contemporary era is an era of 'time–space compression', entailed by modern technology that rapidly transmits information and population. This rapid transmission leads not only to the world-wide connections between places and their dynamics, but also transforms their characteristics. In this context, places are becoming more complex than ever. Massey (1994: 146–156) then argues for 'a global sense of place'. Massey (1994: 155, 156) suggests that the uniqueness of place derives from various sources, rather than from some long, internalised history alone. Similarly, Frampton (1983) suggests the possibility of cross-fertilisation between a universal civilisation and a rooted culture, between the modern technology and local idiosyncrasy. Frampton (1983: 162) refers this tendency in architectural practice to critical regionalism, stressing 'a commitment to place rather than space'.

Following the above theoretical trends, there have been claims for architecture to pursue global eco-culture or hybrid world-culture (Abel 1997; Frampton 2005; Umbach and Huppauf 2005), with the notion of hybrid culture as sources of design inspiration (Missingham and Selenitsch 2002; Smith 1994), and a dialectic attitude towards values of insiders and outsiders in creating place (Seddon 1997; Corner 1999: 12; Girot 1999). These theories suggest that globalisation and world-wide connections may not simply entail homogenisation, but may sustain a place through cross-fertilisation between the inherent and the new.

On the other hand, international landscape architectural practice can be regarded as one of the many threads of global connection. The flow of planning and design services across the globe is related to the global circulation of capital and culture, facilitated by rapid transmission of information, technology, and skilled landscape architects. In other words, international landscape architectural practice can be regarded as an expression of globalisation. Although the abovementioned theories and discourses suggest a focus on discerning local distinctiveness in the midst of global connectedness, there has also been a lack of consideration of how professional practice itself operates as a form of global connection. Furthermore, little has been done to explore the consequence of such global connection on the process and outcomes of changing places and creating landscapes across cultures.

Recent studies of design practice reveal complicated issues emerging from the cross-cultural context (Lorch and Cole 2003; Bull and Boontharm 2007). These issues often centre on western design practices in non-western societies, on the understanding of local context, expectation and delivery process, and on effective ways to exchange knowledge across cultures. Different practitioners may deal with these in highly intuitive ways. The delivery of planning and design across cultures and on the ground, and the identification and interpretation of local characteristics in planning and design, presents a rich and highly complex field of study.

Specific and in-depth research on international landscape architectural practices, particularly in China, has not been identified in this research to date. The lack of in-depth research contrasts with the fact that international practice in landscape architecture is occurring around the world and is rapidly increasing, especially in China. Cantor (1997) saw international practice as a trend in landscape architecture in the late 1980s and early 1990s. He perceived the expansion of international practice in landscape architecture as facilitated by instantaneous communication and rapid transportation. However, the characteristics of landscape architectural practice in the international context remain unclear. The brief account of practitioners' approaches to place and practice by Cantor (1997: 141–184) hardly addressed the likely complexity of international design practice.

1.3 Aim, Objectives and Scope

The aim of this research was to explore the characteristics of international landscape architectural practice. The first objective of this research was to use theory to examine the approach in landscape architecture to place and practice, the realm and position of the profession in an international context and challenges to international design practice of knowing and re-forming place. This objective led to a review of literature and the building of a theoretical framework for this research. The second objective was to investigate the characteristics and processes of international landscape architectural practice in a specific and real-life context. This objective led to a review of key texts about the chosen context, a survey of international landscape architectural practice and a case study in this specific context.

To achieve the above objectives within this research, a clear scope was required. First, the study focused on international design practice between different cultures (i.e., Western and non-Western cultures). This was not a study of cross-national design practice within relatively similar cultures, such as American firms providing planning or design work for Australia and the UK (West to West), or Japan-based firms for Malaysia and China (East to East). This kind of international practice deserves its own research attention and is therefore excluded from this research.

Second, the study refers to Australia as supplier of international landscape architectural services, and China as a site of demand. These references are used for three reasons. First, the research was conducted mainly at the University of Melbourne, Australia, providing access to Australian landscape architects and landscape architectural businesses practicing internationally. The second reason is that China has already been a major site of international practice in landscape architecture and is a key player in the Asia–Pacific region. Therefore, to a certain extent, Australian landscape architectural practice in China may reflect a larger cross-cultural divide between the West and non-West. The third reason is that my background is Chinese. I have lived in Australia for more than five years. I have already been informed about international practice in landscape architecture through my professional experience as a landscape architect in both Australia and China. This socio-cultural familiarity can also facilitate any study involving both countries.

Third, this research focused on landscape architects in the context of international planning and design practice. Landscape architectural practice may involve not only landscape architects but also other professions, such as urban planners, urban designers, artists and architects. Landscape architectural services can also be provided by multidisciplinary firms rather than landscape architectural firms alone. This definition was set to ensure the focus of the profession of landscape architecture.

Lastly, the sources of information used to explore international landscape architectural practice in this research included books, planning and design reports, and journals. Most of these were in English, though some of the more recent professional journals from China appear as citations in English with their Chinese titles in the bibliography. Another important repository of data was created through a survey of Australian landscape architectural businesses, interviews and field work relating to the chosen case study: the planning and design of the Li Lake area. These included questionnaires, transcripts, photographs, plans, brochures, and a small number of personal communications. A detailed explanation of the methods adopted for this study, the sources of data, and the approaches to collect and analyse data is provided in Chapter Three: Research Design.

References

Abel, C. (1997). *Architecture and identity: towards a global eco-culture*. Architectural Press: Oxford, [England]; Boston.

Bull, C., Boontharm, D., et al. (Eds.). (2007). *Cross-cultural urban design: Global or local practice?* London.

Cantor, S. L. (1997). *Contemporary trends in landscape architecture*. Wiley, New York.

Corner, J. (Ed.). (1999). *Recovering landscape: Essays in contemporary landscape architecture*. New York.

Frampton, K. (1983). *Seven characteristics of critical regionalsim*.

Frampton, K. (2005). Critical regionalism revisited: reflections on the mediatory potential of built form. In M. Umbach & B. HÜppauf (Ed.), *Vernacular modernism: heimat, globalization, and the built environment* (pp. 193–198). Stanford, Calif., Stanford University Press.

Girot, C. (1999). Four trace concepts in landscape architecture. *Recovering landscape: Essays in contemporary landscape architecture* (pp. 59–67). Princeton Architectural Press: New York.

Harvey, D. (1990). *The condition of postmodernity: An enquiry into the origins of cultural change*. Blackwell: Oxford [England]; Cambridge, MA, USA.

Lorch, R., & Cole, R. (Eds.). (2003). *Buildings, culture and environment: Informing local and global practices*. Malden, MA, Blackwell Pub.

Massey, D. B. (1994). *Space, place and gender*. Cambridge, Polity.

Massey, D. B., & Jess, P. (Eds.). (1995). *A Place in the world?: Places, cultures and globalization*. Oxford, England; New York, Open University; Oxford University Press.

Missingham, G., & Selenitsch, A. (2002). *Transplants, transferences, translations: Exchanges of strategies and design ideas, garden and cross-cultural considerations* (pp. 1–18).

Museum of Dong-shan-sheng. (1931). *The pictorial record of Chang-chun garden*. University of Dong-bei.

Norberg-Schulz, C. (1980). *GenMius loci: Towards a phenomenology of architecture*. Academy Editions.

Relph, E. C. (1976). *Place and placelessness*. Pion.

Seddon, G. (1972). *Sense of place: A response to an environment, the Swan coastal plain Western Australia.* W.A., University of Western Australia Press.
Seddon, G. (1997). *Landprints: Reflections on place and landscape.* Cambridge University Press.
Smith, K. (1994). Relics, Prosthetics and Surrogate Realities. *The Culture Of Landscape Architecture,* 211–221.
Tuan, Y.-F. (1977). *Space and place: The perspective of experience.* University of Minnesota Press.
Umbach, M., & HÜppauf, B. (Eds.). (2005). *Vernacular modernism: Heimat, globalization, and the built environment*. Stanford University Press, Stanford, CA.
Wang, Y. (2008). Study on formation mechanism and characteristics of early parks and open spaces along the bund. *Journal of Shanghai Jiaotong University (agricultural Science), 26*(2), 91–95.

Chapter 2
Knowing and Re-forming Place

Abstract Chapter Two explores approaches to knowing and re-forming place that has been theorised in the discipline of landscape architecture and discussed in relation to international design practice. This chapter provides a theoretical framework for the study of the practical work of international landscape architectural practice. Chapter Two has three major sections. Section One discusses landscape architecture's approach to place and practice and its primary goal to sustain the characteristics of place. Section Two analyses a notable realm for contemporary international design practice, in which Western designers are providing services for Asia's urbanisation, including landscape architectural services. Section Three discusses the challenges of knowing and re-forming place, presented in the context of international design practice.

Keywords Local understanding · Landscape architecture's approach · International practice · Cultural difference · Urbanisantion

2.1 Landscape Architecture: Approaches to Place and Practice

The discourse regarding contemporary landscape architecture suggests that landscape architecture's approaches to place and practice have a primary goal to sustain the characteristics of place. Corner (1999: 12) claims that landscape architecture has traditionally sought to recover sites and places. Further, Corner (1999: 13) argues that later modernist architectural and urban planning and design overlooked 'local characteristics and values', resulting in the failure of universal and utopian trends. Corner (1999: 13) argues that 'landscape is seen as a means to resist the homogenization of the environment while also heightening local attributes and a collective sense of place'. Likewise, Fieldhouse and Harvey (1992: 9) argue that retaining and restoring a sense of place are 'established tenets' of landscape architecture that distinguish it from other creative disciplines.

This primary goal to sustain the characteristics of place is evident in a number of ways. One is the value the profession assigns to place components that are not easy to

Y. Zhang, *Place Making in International Practice of Landscape Architecture*, Advanced Topics in Science and Technology in China 64,
https://doi.org/10.1007/978-981-16-2442-1_2

notice, see, or find (i.e. hidden aspects of place) and the emphasis on gaining understanding of them. Another is the expansion of the scope of professional knowledge regarding place in the urban practice of landscape architecture as a response to the pressures of urbanisation. These two threads also suggest landscape architecture's approach to using history to inform the newly urbanised landscape and the cultural construction of place. These ideas are discussed further below.

2.1.1 Valuing Hidden Aspects of Place

Hidden aspects of place refer to those that are often difficult to define or interpret but crucial to the quality of place. According to Girot (1999: 63), the French landscape architect and educator, hidden aspects of place are '*je ne sais quoi*' ingredients that convey distinct qualities. Hidden aspects of place also exist in a multitude of forms, from non-visual or sensorial aspects of place, to the meaning attached by cultures to seemingly normal or banal elements. Rogers (2001: 21), the American landscape designer and landscape preservationist, argues that a physical place 'mirrors the human mind as a theatre of myth, ritual, allegory, and reason'. This suggests that hidden aspects of place comprise spiritual, experiential and religious ingredients. In short, hidden aspects concern the meaning of place.

Compared to physical entities and visible appearances, hidden aspects of place are easily overlooked. Paterson and Pettus (1994: 139), the Australian landscape architects, suggest that the approach to revealing the true quality of place is a site-specific approach. They (1994: 139) argue that such an approach can correct 'the repetitive use of catalogue details' and 'functional problem-solving' that lead to the production of a 'monotonous and homogenous non-place'. The above arguments suggest that first, hidden aspects convey the true quality of a particular place and are thus important; second, the profession of landscape architecture values these aspects and has developed ways to gain knowledge of them; and third, the purpose of revealing and valuing hidden aspects is to sustain the characteristics of a particular place.

Paterson and Pettus (1994: 140) explain that the site-specific approach can inspire landscape design in four ways. The first is by engaging with 'participant[s] and/or user[s] both physically and intellectually' to make the landscape visible and no longer mere background (1994: 140). The second is through studying the language of landscape forms for a given site. This is for understanding 'a complex series of associations and meanings which are prescribed by the users and participants' (1994: 140). The third is the understanding of natural systems and the human ecosystem. Paterson and Pettus (1994: 140) argue that the order of nature and natural systems are the most powerful and expressive sources of inspiration for landscape architecture, suggesting that natural systems and the human ecosystem should be viewed together and reconciled. The fourth method to inspire design is the study of history to define the culturally relevant and significant historical forms in the context of a site.

The four design inspirations of a site-specific approach suggest that in the vision of contemporary landscape architecture, hidden aspects of place are considered in two ways. One is the meaning of place and the other is the broader historical, socio-cultural and natural context. Both are less visible or tangible. Second, these aspects are understood by engaging with the people who use or change the place and by studying a site as a cultural landscape and a balancing system. This leads to research into the history of a site or place. In terms of design process, gaining such knowledge of place is essentially a research stage of landscape architectural practice, aimed at familiarising the landscape architect with a site or place to constructively propose changes. That type of research involves a process of immersion in a site or place.

Given that understanding, it should not be surprising that design scholars in landscape architecture emphasise the strategies for deeply exploring a site or place. In his article, *Four Trace Concepts in Landscape Architecture*, Girot (1999) argues for a strategy of landscape investigation and design, including 'landing', 'grounding', 'finding' and 'founding'. Girot (1999: 61) describes landing as the first step in moving from the unknown to the known, requiring reactions to the difference between the preconceived notions of designers and their experience of the real site. Therefore, landing has 'potent spatial and psychological effects on the subsequent thinking-through of the design project' (1999: 62). While landing describes the personal experience of the landscape architect, the second stage, grounding, moves from personal thought to the objective discovery of significant site elements. This process involves careful research and analysis of the site's surrounding context (its soil, climate, water, ecology and history); an objective probing into its successive histories; and an exploration of the forces and events underpinning the evolution of place (1999: 62–63). Girot (1999: 63) further argues that these forces and events are more important to understand than those that are visible to the eye. Such an argument suggests that hidden aspects of place comprise past events, which are understood by research into history.

Girot's argument for landing and grounding also suggests that the initial experience of designers (landing) and their on-site investigations and research of the history of the site (grounding) are important for knowing places. These two steps are also important for identifying differences between the preconception of designers and the reality of a site and for forming initial design concepts. Compared to Paterson and Pettus's site-specific approach, Girot's theory stresses that the personal experiences of designers are essential for knowing places. This initial experience needs the support of careful site investigation and further research into the history of a site or place.

Following the stages of landing and grounding, Girot argues that the next stage, 'finding', is an 'alchemical' process that results from the first two stages and includes the intuitions of designers during landing and the subsequent discovery of the site during grounding (1999: 63–64):

> What is found is the *je ne sais quoi* ingredient that conveys a distinct quality to a place. As such, findings escape design invention and import; they are something unique (though hidden) that definitely belongs [belong] to a place and contributes durably to its identity [...] Finding usually discloses the evidence that supports one's intuitions about a place.

Girot (1999: 64–65) describes the final and fourth step, 'founding', as a synthesis of the prior three steps and the most durable and significant experience among the four. Although the act of founding can either be conservative or innovative, Girot (1999: 64–65) argues that:

> [...] the act of founding is always a reaction to something that was already there [...] the act of founding corresponds[...]to a given period in history when a cultural relationship to the landscape evolves and changes [...] a well-founded project remains clear in its approach and resolution, extending the legacy of a place toward a productive future.

Girot's arguments for finding and founding suggest that those concepts underpin the process of generating design ideas and that design ideas are based upon the discovered distinctiveness of place. The argument also suggests that research into the history of a particular place plays a crucial role in landscape architectural practice. Historical research generates a deeper understanding of hidden aspects of a place, especially forces and events that underpin its transformation, and leads to solutions for sustaining the characteristics of the place during its development.

Girot (1999: 60) considers that the above four trace concepts underline the fact that professional designers seldom belong to the place where they are asked to intervene. Such a view parallels that of Corner (1999: 12), who argues that designers are always outsiders, 'as far as the beholding of manufactured landscape goes, for to be inside entails the evaporation of landscape into everyday life or milieu'. These viewpoints that designers are outsiders further explain the particular concern of the profession for hidden aspects of place that include the cultural context and natural systems of a particular place. They also explain the emphasis of theorists on ways to really know places, which include designers' personal experience of place, their careful on-site investigations, personal engagement with people, and research into the history of a particular place.

It could be argued that the term 'sense of place' has been popularly used in landscape architectural discourse to refer to something distinctive but not easy to identify, interpret or represent. In landscape architectural practice, the process of knowing hidden aspects of place is a process of gaining implicit knowledge covering a particular place. Although knowing explicit aspects of place, such as physical features and environmental statistics, is also necessary, hidden aspects convey true quality, thus, are more critical to the characteristics of place, although they are often overlooked. This problem underlies landscape architecture's approach to place and practice and its primary goal to sustain the characteristics of place.

2.1.2 Expanding the Scope of Place in Urban Practice

The discourse of contemporary landscape architecture has suggested that the profession is expanding the scope of professional knowledge from the site or place to the context—the landscape setting accommodating the site. This shift can be interpreted as a response to the pressures of contemporary urbanisation and is evident in arguments that seek urban identity in nature and ecology, by emphasising nature conservation and expertise in ecology. This shift also suggests that the working territory of the profession of landscape architecture is expanding from design to planning.

Urban development is a process of re-forming places. This process, however, implies homogenising effects that may lead to the loss of the characteristics of place. In his book, *Out of Place: Restoring Identity to the Regional Landscape*, Michael Hough (1990: 87), the Canadian educator, professor and practitioner in landscape architecture, who was mentored by Ian McHarg, argues that, before the explosion of cities, towns drew their character from their regional settings. Being inside looking out or outside looking in often caused a sense of context upon which one's whereabouts were known (1990: 88). That sense of context has decreased as the growth of contemporary cities has fragmented landscapes (1990: 88). Hough (1990: 88) argues that 'where the land has retained its basic topographic or biological character, either fortuitously or by design, it can maintain something of its original identity'. Hough (1990: 116) further proposes that the preparation of a landscape conservation strategy preceding urban growth may reduce much of the placelessness of new developments. Hough's argument suggests nature conservation as a basic countermeasure in re-forming places as a response to the homogenising effect of urbanisation.

Hough further provides six principles for 'regional design'. These principles suggest that the scope of professional knowledge of the characteristics of place has expanded to the identity of landscape, the working territory of landscape architecture from design to planning, and an emphasis on knowing the interrelated natural and social forces that change places throughout history. The six principles include 'knowing the place'; 'maintaining a sense of history'; 'environmental learning and direct experience'; 'doing as little as possible'; 'sustainability' and 'starting where it's easiest' (1990: 179–195). Each principle is based upon Hough's idea that the identity of the regional landscape arises from the character of change that occurs there over time. In Hough's view, the interactions of communities with their land and natural environment initially shaped the identity of the regional landscape. In this process, changes were driven by practical needs. In the present day, choices made by planners and designers substantially influence the identity of regional landscape (1990). Hough (1990: 195) argues that:

> The sense of place is organic, changing with time. As new human needs and technologies exert themselves on the landscape, new landscapes emerge in response to the imperatives of changing conditions.

Of the first principle, 'knowing the place', Hough suggests that what one needs to know about a place in order to propose change through planning and design includes 'the natural process of the region or locality' and the social process through 'the collective reaction of people to the environment over time' (Hough 1990: 180). For the second principle, 'maintaining a sense of history', Hough (1990: 186) includes the history of both nature and culture that lies in 'the historic interactions of man and nature'. For the third principle, environmental learning and direct experience of planners or designers, Hough indicates the importance of understanding the everyday lives of local people in ordinary and often unnoticed, but memorable, places. The importance of understanding how communities live in a particular place shows clearly in this third principle.

Hough (1990: 190–191) also claims that the greatest diversity and identity of places often results from a minimum of outside interference. This fourth principle, doing as little as possible, is considered the antithesis of design style. Hough (1990: 193) argues that a design style implies characteristics by which a designer is identified rather than the designed landscape or place. He argues for four methods to assist designers to understand and design a place while imposing a minimum of change. First, designers or planners understand the processes that make things work. Second, they propose planning and design strategies that enable the development of natural or social diversity. Third, they know where to intervene. Fourth, they propose planning and design strategies that enable natural diversity to evolve on its own. The principle of 'doing as little as possible' suggests imposing minimum outside forces on a desired development. Such an aim suggests that planners and designers should do as much as possible to identify the underpinning characteristics of place.

In his fifth principle, sustainability (Hough 1990: 193–194), Hough argues for the potentially positive role of both development and technology in sustaining place, if development and technology are properly positioned in planning and design. Hough considers this principle as a common ground between nature conservation and developmental initiatives, which finally relates to environmental ethics. Hough (1990: 194–195) describes the sixth principle, 'starting where it's easiest', as 'an encouraging environmental principle to follow in bringing about change'. Hough argues that this last principle is 'the only practical basis' that integrates technology and 'people, urbanism, and nature in ways that are biologically and socially self-sustaining and mutually supportive of life systems'.

Hough (1990: 179) relates these six principles with what he called 'regional design'. This association suggests the intricate relationship between the work of design and planning in landscape architectural practice, as well as between the characteristics of place and the identity of landscape. In Hough's (1990: 15) words, 'landscape is the expression of place's regional context especially in the absence of distinguishing architectural styles'. In the words of Relph (1976: 123), 'landscape is both the context for places and an attribute of places'. These arguments suggest that the characteristics of a place interweave with the identity of its regional setting. Therefore, knowing a place involves knowing its regional setting. In other words, for landscape architects, the scope of professional knowledge of a particular place is broader than that of the presented site or place.

Hough's six principles again suggest the need to know three aspects of place in landscape architectural practice. These include first, the history of place; second, local people and their everyday life, especially those in unnoticed but memorable places; and third, the natural and social systems underpinning the evolution of places. As argued in the previous section, these are hidden aspects of place. Comparing Paterson and Pettus's (1994) site-specific approach and Girot's four trace concepts, Hough's principles more clearly show the significance of gaining knowledge of these hidden aspects through direct engagement between planners or designers and local communities, and through extensive research into the history of the evolution of sites or places. Further, the third aspect for knowing, the natural and social systems of a place, is clearly aimed at resisting the homogenising effect of urban development.

The expanding scope of what site or place means to the profession of landscape architecture and their continuing interest in hidden aspects of place are further evident in recent arguments for landscape urbanism focusing on ecology. Shannon began her argument in her article *From theory to resistance: landscape urbanism in Europe* (2006: 141) with a discussion of critical regionalism, which, she argues, implies environmental determinism. According to Shannon, such environmental determinism has recently 'extended to critical regionalism in the medium of landscape, in an effort to challenge the internationally imposed generic models of modernisation and urbanisation and to resist the homogenising effects of late capitalism' (2006: 141). Shannon (2006: 144) argues that 'site-based landscape' is assumed to be an intermediary between built form and an otherwise placeless urban surface. Shannon (2006: 145) further claims that landscape architects are currently positioned at the 'convergence of agrarian peasantry (low culture) and urban design (high culture) heritage'. Shannon (2006: 145) suggests that the study of landscapes by landscape architects should begin with the 'recollection of previous history' and then tends to anticipate 'future re(colonised) sites'. Such approaches suggest that the possible resistance to placelessness through the urban practice of landscape architecture is based upon knowledge of the history of the landscape setting of a particular place, including both of the rural and urban history.

Shannon (2006: 146) suggests that landscape urbanism is promoted by the contemporary condition of uncertainty. She (2006: 146) connects this uncertainty to 'indeterminacy regarding future development and the incapacity to shape it is [as] a definitive form', suggesting discussion of landscape beyond form, tangibility and objective quality. This formlessness, intangibility and non-objective quality, as shown in Shannon's argument, suggest values placed by the profession on hidden aspects of place yet from a broader perspective. According to Shannon (2006: 146), these hidden aspects are 'the evident foundations of the physical and geographical logic expressed by the site', which lead to the practice of landscape urbanism that focuses on infrastructure, ecology and 'post-industrial natures' (2006: 146–152).

Shannon's (2006) argument for 'site-based landscape' recalls Corner's (1999) proposition for expanding the scope of the site-specific design approach. Corner (1999: 12–13) argues that the traditional approach, the employment of 'site phenomena as generative devices for new forms and programs', and the assumed 'mnemonic and temporal significance' are renewed by the recovery of 'biological importance'. Corner (1999: 13) then outlines three possible measures to reclaim sites and places in present-day landscape architectural practice. The first is 'the retrieval of memory and the cultural enrichment of place and time'. The second is the 'social program and utilities' for new uses and activities. The third is 'ecological diversification and succession'. The final measure through ecology is also evident in Hough's (1990) early argument for regional design. Hough shows that an understanding of the interrelated natural and social forces driving place evolution can support minimum intervention in a desired development. It can support the continuity of the characteristics of place. Similarly, Paterson and Pettus' (1994) site-specific approach suggests the need to view natural systems and human ecosystems together. These arguments all suggest the understanding of the balancing system of a particular place (i.e., its ecology). In relation to Corner's (1999) argument for ecology, it could be argued that the emphasis by theorists on ecology became increasingly clear during the 1990s, coinciding with the development of the urban practice of landscape architecture.

The emphasis on ecology parallels reheated topics of meanings and appearances of the landscape accommodating sites and places. Czerniak (2006: 122) argues that while the trend of the late 1990s emphasised landscape as agent, verb and process, the present landscape architectural practice needs a full engagement with 'the etymology of landscape' rather than only favouring 'time over space, performance over appearance, effect over meaning'. Czerniak (2006: 121) further claims that 'to think site is to think landscape'.

It could be argued that Corner's (1999: 13) other two propositions regarding the measure of cultural enrichment and new activities for recovering landscape suggest a broader perspective on the scope of the meaning of place in the urbanising landscape. This broader perspective is also evident in the argument of Smith (1994: 212), the American landscape architect and educator, who claims that 'the fragmentation of urbanity' and the 'hybridization of cultural influences' provides multiple sources of inspiration that influence the way we conceive and design landscape. Such a claim aligns with the idea that the practice of urban design, including landscape architecture, is inevitably cross-cultural (Bull and Boontharm et al. 2007). In this sense, in the urban practice of landscape architecture, professional knowledge should extend not only from art creation to ecological recovery, but also from local or regional culture to contemporary world-culture.

The above discussion suggests the expanding realm of the knowledge of how to constructively re-form places, especially under pressures from urbanisation. New visions of the characteristics of place in the urbanising world can be assumed in present-day landscape architectural practice. First, the natural environment and its biophysical characteristics are the means by which contemporary urban places and cities can define their characteristics. Second, approaches to knowing places and their landscape settings emphasise local ecology as a balancing system, involving the understanding of both the natural and social systems of the site-based landscape. Finally, in the vision of the profession of landscape architecture, the characteristics of place sustained in an urbanising world have new ingredients generated from contemporary cross-cultural relations, and can be expressed in new ways rather than more traditional concrete forms.

2.1.3 Using History to Inform the Newly Urbanised Landscape

Assuming that the profession of landscape architecture is capable of recognising hidden aspects of place, identifying their values, taking a broader perspective, and re-forming places based upon their natural and cultural legacies, history should play a key role. It could be argued that landscape architecture's approach to using history to inform the cultural construction of place, especially in the newly urbanised landscape, differs from the use of history in existing cities. Of the latter, historical buildings, streets, and monuments form the material evidence of the history of a particular place. In landscapes experiencing a process of rural-to-urban transformation, similar evidence of cultural values attributed to the previous rural environment, such as bridges, wells or cottages, are often physically decayed along with the relocation or changes to life of the original peasant community. Therefore, in the practice for rural-to-urban transformation, landscape architects require a range of methods to conduct historical research.

In urban planning, historical research for revealing values of place shows a trajectory of concerns, from architectural forms to formless collective memory. The latter suggests ways of using history to inform the cultural construction of place in the newly urbanised landscape. In the book (Monclus and Guardia 2006), *Culture, Urbanism and Planning*, Hebbert and Sonne (2006) explain this trajectory using three historiographic strategies that they argued characterise twentieth-century town planning in the West. In the first strategy, history provides an 'encyclopaedic source' for generic types and forms of space (2006: 4–10). In the second, history provides 'the reservoir of collective memory, place-identity, and local attachment' (2006: 4, 11–13). In the third, history provides 'a source of normative justification and purpose for the act of planning' (2006: 4, 13–18). Hebbert and Sonne (2006: 4) argue that each strategy implies a distinct pattern of time and way of reading the past and that each strategy has its own risks, suggesting the complexity of using history and tradition as a reference in re-forming places. Their three strategies for the use of history in town planning are further explained below.

Hebbert and Sonne (2006: 4) argue that the first strategy (i.e., history as lasting architectural types and forms) is essential to guarantee the survival and continuity of types and forms, including 'lasting components and attributes'. They argue that the second strategy (i.e., history as local memory) itself has three phases of influential thoughts and methods. The first phase is the method of 'Survey Before Plan', using 'intensive historically based local survey as a base for action' (2006: 11), which was originally proposed by Scottish town planning activist and theoretician Patrick Geddes at the beginning of the twentieth century. The second phase completely reorients Geddes's idea as applied by modernist town planners in the mid-twentieth century, collecting facts merely to define 'an empirical basis for technical action' (2006: 12). The third phase is 'a broad rediscovery of the local dimension', typified by the impetus of the conservation movement and desire to protect historical settlements and monuments in the Anglo-Saxon world in the 1960s and 1970s (2006:

12). Hebbert and Sonne regard the last type of historiographic strategy as 'a Big Canvas' (2006: 13), whereby the methods of documentation, oral history and public participation for discerning the collective local memory are combined with earlier methods. Further, Hebbert and Sonne (2006: 19) point out that for the first time in fifty years, urbanists now see themselves as environmental stewards, rather than as only agents of modernity and technical progress.

The above historiographic strategy suggests that in addition to the study of lasting architectural forms, the study of collective local memory is an important approach to understanding a place. Such understanding can be gained through the collection and study of documents, oral history and public opinion. This strategy suggests methods for using the non-material evidence of history to inform the cultural construction of place in the newly urbanised landscape, which is through engagement with the peasant community.

Further, while the role of urban planners in environmental stewardship is relatively new, it has long been a clearly defined, central task of the profession of landscape architecture. This fact suggests that the expertise of landscape architects can help ensure environmental stewardship in urban development. It could be argued that such environmental stewardship comprises the work of engaging with aboriginal communities, which is influential to the cultural construction and ecological recovery of place. In relation to Wall's (1999) argument that urbanising peripheral areas are of particular significance to contemporary planning and design, it could even be argued that landscape architects are playing an active role in the urban transformation of these areas, facilitated by their position in environmental stewardship.

In terms of cultural construction of place, however, strategies for planning and design are increasingly confined by strategies for economic development. Seddon's (1997) earlier argument against tourism and the capitalisation of local characteristics, which often results in the destruction of landscapes, displays that concern. In the last decade, economic motivation and its consequent cultural strategies have, by contrast, been described as protecting and sustaining local growth within the global economy. Monclus and Guardia (2006: xv–xvi) argue that since the 1980s, the term culture 'has become converted into the premium material of the urban economy and has been seen to be invaded by economic logic and the practices of marketing' (2006: xiv). Further, they (2006: xviii) argue that the growing intensity of urban competition in the global market promotes the integration of culture and economy, and that integration promotes the convergence of two previously divergent trends in urban planning. Of these divergent trends, one trend shows a progressive vision: looking to the future, acting as an agent of modernisation and emphasising approaches through science and technology (Monclus and Guardia 2006: xv). The other trend is culturalist planning: looking to the past and assuming forms of nostalgia whilst stressing the values of urban community and historical cities over modern banality. Monclus and Guardia's argument suggests that in contemporary urban development, first, cultural ambitions are likely to be implemented by new technology; and second, ways to use history are likely to be directed by the pragmatism of economics.

It could be argued that in such an alliance, ways to use history to inform the newly urbanised landscape can also be directed by developmental efficiency, economic purpose and advanced technology. This may, in turn, emphasise image-building in urban landscape. Examples of image-building can be the picturesqueness of places and the use of labels as an instant way to express the historical values of places and sites. As discussed previously, in the peripheral areas of cities, where contemporary rural-to-urban transformation is occurring, the cultural construction of place is informed most significantly by the non-material component of history, such as events and a collective local memory of the original peasant community. While economies drive cultural strategy in urban planning, similar economic purposes can be assumed in the urbanising landscape. This suggests that these peripheral areas are practically useful for observing ways of landscape architects to research into history, identify values, and use them creatively for the cultural construction of place.

The need to carefully research the non-material evidence of history and the present economic purpose driving the transformation of rural landscapes into urban places suggest that there is a substantial need for work from landscape architects throughout the research stages of planning and design practices. The research must cover the cultural, historical context and ecosystem of sites and places. It could be argued that landscape architectural practice in the international domain demands reform and increasing emphasis on research and sufficient understandings of place. In China, for example, rapid and vast urban development is occurring, inducing demand for international landscape architectural services. However, tight schedules for development leave little time for international practitioners to 'land' and 'ground', in Girot's terms; and to conduct comprehensive historical research—the 'big canvas', in Hebbert and Sonne's terms. The next section discusses a dominant realm of international design practice—Western practice in Asia's urbanisation, in which a potentially prominent position can be predicted for landscape architecture.

2.2 International Landscape Architectural Practice: Realm and position

A notable field for international design practice is in Asia. In this field, planners and designers from developed Western countries are providing services including landscape architectural services for the urban development of Asian countries. Such international planning and design practice is also cross-cultural. In this cross-cultural realm of international design practice, a prominent position for landscape architecture can be assumed because landscape architects' disciplinary-specific skills appear to facilitate their knowing and constructively re-forming place across cultures.

2.2.1 *Western Design Practice in Asia's Urbanisation*

The participation of Western design practice in Asia's urbanisation is evident in Western architectural practice there. The world survey of architectural firms in 1997 found that thirteen of the largest thirty firms operating in Asia were based in the United States, Australia and the United Kingdom (Marshall 2003: 2), which are all dominated by Western cultures. Marshall raises questions about such a phenomenon, asking whether international design practice is just 'a case of foreign firms exploiting the Asian situation' or reflects 'Asian clients demanding an international aesthetic' (2003: 2). The Asian situation and the demand for an international aesthetic by Asian clients can be further related to the vast scale of urbanisation occurring in Asian countries such as China and the historical asymmetry caused by Western colonisation in Asian history.

First, Asia's urbanisation has different conditions from the West. Hall and Pfeiffer (2000: 4) suggest that between 2000 and 2020, the percentage of the urban population in developing countries, comprising Asia, Africa and Latin America, will rise from 41 to 52%, while in developed countries, it will rise from 76 to 81%. In developed countries, 'the great period of urbanisation has already been and gone' (2000: 3–4), and the urban transformation of Asia, as well as Africa, has become more influential in world urbanisation. The authors exclude Latin America because its urban population has already reached almost 75%, while that of Asia and Africa remain less than 33% (Dutt 2003: 4).

Moreover, Asia is distinguished from Africa by its greater 'magnitudes' of urbanisation (Ness and Talwar et al. 2005: 9). It is expected that by 2030, Asia will have 58% of the world's urban population and Africa will have less than 16% (2005: 8–9). Likewise, Neilson et al. (2000: 5–9) argues that the growth of the global population and large-scale urbanisation, especially in Asia, will result in massive social transformation and have implications around the world. Such discussions reveal that the twenty-first century will see most urbanisation occurring in the developing Asian countries at a scale much greater than that of developed Western countries.

Further, there are two different patterns and forces driving urbanisation that characterise developed and developing countries. From their global perspectives, Pomeroy and Webster (2008: 11–12) argue that the process of 'rural-to-urban' migration is often 'productionism oriented', whereas the 'urban-to-rural' migration process is often 'environmentalism driven'. Both are observable in developed countries. However, productionism-driven urbanisation is dominant in developing countries, whereas environmentalism-driven urbanisation is becoming more prominent in developed countries (2008: 11).

Second, the so-called international aesthetic requested by Asian clients for new urban development implies a willingness to catch up with the West and to promote ideas of a competitive Asian economy. Using China as an example, Zhu (2004: 14) points out that there is a 'historical asymmetry' existing in China in relation to Western knowledge. By historical asymmetry, Zhu means that in the period from the eighteenth century to the twentieth century, China's modernisation was a process of transplanting Western knowledge, following Western models and lagging behind those models. It could be argued that this historical asymmetry also applies to many other Asian countries because of their colonial history and the consequent one-way transfer of technology and theories, lasting for nearly two hundred years. It even could be argued that such a tradition of historical asymmetry encourages Asian clients to invite Western planning and design practice as an extension of this generalised trend.

Although Chinese classical gardens, handicrafts and interior decorations have continued to attract the West since the eighteenth century (coined as the taste of chinoiserie), this does not deny contemporary cultural asymmetry. The contemporary practice of Chinese-style gardening in western countries remains, such as the practice of kitchen gardens by Chinese immigrants in Australia (Armstrong 1999) and the Chinese garden in Sydney's Darling Harbour designed by Suzhou Landscape Design Institute in China. These examples suggest individual practices or the export of landscape architectural services from China as part of 'sister city arrangements' (Missingham 2007: 118), in other words, as a form of diplomacy. These sporadical examples do not suggest that exporting design services from China to western countries is main stream, but in turn confirms the historical asymmetry manifest in providing professional planning and design services.

The above discussions suggest that Western design practice in Asia dominates and the process is rarely vice versa. This is also evident in the contrasting urban reality between the developed West and the developing Asian countries. On the one hand, there is less urban development in the West compared to its great period of urban expansion in the nineteenth and twentieth centuries. Urban theories and practices are, however, more mature than in the non-West. On the other hand, in non-Western countries and in Asia in particular there is demand both for design practice and for theories to underpin the new forms of urban development. Such a discrepancy results in Western design firms practicing in Asia's urbanisation.

This cross-cultural realm of international design practice also suggests opportunities of cross-fertilisation through practice. Simmonds and Hack (2000: 4) claim that the end of the twentieth century was the start of 'a new urban epoch' where, for the first time, the whole world faced the same forces of change. By the same forces, the authors mean global culture and economy, suggesting that similar forces of change influence both the West and the non-West. Such views suggest that the nuanced difference between Western and Asian urbanisation is also embedded in the shared global forces of urbanisation, as was also indicated by Bull and Boontharm et al. (2007). Therefore, international design practice could be also regarded as the result of the shared global forces of change, not just the different urban conditions between the West and the non-West. The shared forces of urbanisation—global culture and economy, may not only promote design practice as cross-cultural, but may also make reference to global culture. Such conditions suggest the importance of cross-cultural planning and design practice as a process of cross-influence and cross-fertilisation.

It terms of cross-fertilisation, the urban frontier of Asian countries seems to provide Western planners and designers with materials for planning and design thinking. Simmonds and Hack (2000: 4) argue that in the present urban conditions in the West, urban practitioners remain uncertain of their current roles and future directions:

> We live in a 'post-industrial city', with a culture of 'post-modernism' and under the regime of political economy often called 'post-Fordism'. The urbanist Edward Soja has recently coined the phrase "post metropolis" to describe the city region of today. At least, from the point of view of the practitioner, we know what we are not but we do not know what we are or where we are going.

Such doubts about the role and direction of Western urban practitioners, albeit practicing in the West, arise from an analysis of the broader context of the global urban epoch and aim to draw references from emerging urban forms worldwide.

This cross-fertilisation is also apparent in the international architectural design practice in China. Zhu (2005: 479–480) studies a particular professional community in China, including Western practitioners and Western-educated Chinese returnees engaged in architectural design practice and education. He (2005: 479–480) argues that the cross-cultural interaction in such a community engenders 'brutal pragmatism' that may inform the West of its 'lack of recourses to support its own critical architecture'. J. F. Zhu further suggests that Asia's urbanisation, and China's in particular, has created the frontier of urban practice and is challenging Western design ideas and paradigms. Following his argument, Zhu (2006) argues that the experimental and collaborative approaches to research and design are often adopted by the above-mentioned practitioners without a preoccupying design ideology. Such discussions elucidate the positive factors specifically engendered in the cross-cultural domain of design practice rather than cross-cultural challenges per se.

The value of international design practice to the development of local places is also suggested in existing theories. Earlier, Tuan (1977: 18) pointed out that 'a familiar image may lack sharpness unless we can see it from outside and reflect upon our experience'. Although Tuan was looking at the experience of space and how space becomes place through meaningful interaction, his analysis also suggests that outside positions may also engender design intelligence, as unfamiliarity sharpens thinking and promotes reflection.

More recently, King (2007) argues that cross-cultural design practice can be an opportunity. He (2007: xxi) points out that in 'more quarantined, local design practice, hyper-tradition has tended to fall into caricature'; in contrast, in trans-national design practice, there is a perceived compulsion to find the local and to reinterpret it. King uses the term 'hyper-tradition' to describe a local tradition used in design as a means to exaggerate and establish difference between competing cities, such as Bangkok competing with Hong Kong, Singapore and Shanghai. Hyper-tradition is manifest in the 'exaggeration of old practice and images' or 'even the invention of new traditions' (2007: xxi). Such discussions assert the value of international design practice to local places. It suggests that the urban contemplations of today may inform and transform the future more widely through a process of cross-cultural design practice as knowledge exchange.

Simmonds and Hack (2000: 3) also suggest that the urbanising fringes are now expected to act as new anchors for regional development, where new universities, recreational theme parks, and new office and financial centres are often located. The assumption of the urbanising fringe as a new focus for urban development also suggests demand for landscape architectural services there. In the process of building new urban surfaces to extend into rural and natural lands, many works and services provided by Western design firms should be typified by landscape architecture. The next section discusses further the increasingly relevant role of landscape architecture in urban practice, and its potentially prominent position in the international design market.

2.2.2 *The potential of Landscape Architecture*

The discourse of the urban practice of landscape architecture, as previously discussed, portrays an optimistic attitude amongst practitioners and scholars, who assume that landscape architectural projects play a role in resisting the homogenising effects of global development and urbanisation. Such optimism is based on the deployment of ecology because landscape architects employ ecological disciplines, whereas other design professions appear not to, or to know less. This optimism is also based on the skills of integration and collaboration that are increasingly requested from landscape architects.

New visions of contemporary landscape architectural practice have emerged. These new visions are revealed in theories that describe contemporary landscape as neither the antithesis of cities nor as pastoral scenery outside the city (Mathur 1999) but rather as an evolving process synthesising physical and cultural change (Wall 1999). Wall (1999: 235) argues that landscapes are the 'living, connective tissues between increasingly disparate fragments and unforeseen programs'. According to Wall, landscape, as a noun, refers to the larger setting of buildings, urban spaces and activities; as a verb, landscape refers to a process of formation that can be enabled by the practice of landscape architecture.

The argument that the practice of landscape architecture can improve the connective function of landscape suggests the need for new forms of urban practice and a range of skills to accompany those new practices. Wall (1999: 244–246) suggests six 'surface strategies' as productive principles for designing what he describes as the urban surface. The first is 'thickening', aiming at multiplying the number of groundplanes in public places. The second is 'folding', emphasising the interaction of indoor and outdoor spaces. The third is 'new materials' for new uses, such as those of youth culture. The fourth is 'non-programmed use', to introduce diversity and flexibility in the use of public places. The fifth is 'impermanence' to accommodate changing needs and future demands. The sixth is 'movement', focusing on movement corridors as new vessels for collective life. The common thread between these strategies is the goal of integrating contradictory and heterogeneous elements, and facilitating movement and transformation. Again, Wall (1999: 247) argues that the connective principles of ecology are essential to surface strategies because ecology addresses interrelationships between parts and dynamic systems.

Wall's surface strategies target not only physical re-formation but also social and cultural re-formation (1999: 244). Surface strategies also target a synthetic form of creativity and, consequently, require a variety of skills and knowledge in planning and design. Wall (1999: 246) argues that synthetic creativity is based on the integration of the full suite of the traditional disciplines of the built environment: landscape architecture, architecture, urban planning, and engineering. Wall (1999: 247) then suggests that a flexible, reflexive urban fabric and new landscapes requires a hybrid form of practice that synthesises landscape, architectural and urban skills. In relation to landscape architects' knowledge of ecology, Wall (1999: 247) claims that landscape architecture today assumes a more relevant and active role in addressing regional and ecological questions regarding place, time and process that face present-day society. Wall's arguments show that, in the trend towards a hybrid form of urban practice and towards integrated skills, confidence in the capacity of landscape architecture to deal with present-day problems, including the homogenising effect of urbanisation, is increasing. It could be argued that the more prominent role of landscape architecture has been promoted by the pressure of urbanisation and underpinned by the expertise of landscape architects in ecology. Wall (1999: 247) claims that landscape architects are more ready to apply ecology because many of them have studied or been inspired by it.

The requirement for a hybrid form of urban practice demands a synthesis of various design disciplines. The recent discourse in landscape architecture also emphasises a need for synthetic skills and collaboration between professions. Regarding the need to program the urban surface, Corner (2006) argues that there is an existing lack of synthetic skills in landscape architecture. Corner (2006: 28) regards the process of urbanisation as a mechanism for shaping urban relationships and spatial forms. A site or place ought, accordingly, to be understood according to its relationships to other urban elements, and, if that were so, new ways of interpreting urban forms and function would be required. Corner argues that there is a lack of skills that can express such relationships, including those mapped, understood and proposed by designers. Corner (2006: 31-32) writes:

> It is both tragic and ironic that as designers we are all ultimately interested in the density of building but that most who actually accomplish this can only do so through the typically unimaginative and uncritical techniques of design as a service profession. On the other hand, the visionaries, it would seem, are as always provocative and interesting, but their utopias continually evade the problem of an operative strategy.
>
> [...]
>
> I believe that landscape urbanism suggests a reconsideration of the traditional conceptual, representational and operative techniques. The possibilities of vast scale shifts across both time and space, working synoptic maps alongside the intimate recording of local circumstance, comparing cinematic and choreographic techniques to spatial notation, entering the algebraic, digital space of the computer while messing around with paint, clay, and ink, and engaging real estate developers and engineers alongside the highly specialized imaginers and poets of contemporary culture—all these activities and more seem integral to any real and significant practice of synthetic urban projection. But the techniques to address the sheer scope of issues here are desperately lacking.

Corner's argument suggests that the traditional approach to representing a proposed landscape, such as drawings of plans, sections and perspectives, is not enough to interpret the urbanising landscape. Corner further relates the lack of representational skills to problems experienced in relationships between landscape architects and other professions. His argument suggests the emerging emphasis on synthetic skills in landscape architectural practice and that relevant approaches to enhancing such skills should include the collaboration among professionals in a variety of creative disciplines. Similarly, Spens (2007: 10, 11) argues for the potential for a 'digitalisation and reformulation of the site/non-site parameters in the prevailing urban/rural scenario', showing expectations of new methods generated in today's 'laser\net world' and the collaboration between landscape designers, architects, engineers and ecologists.

The emphasis on the need to enhance synthetic skills is also evident in the problematic gap some perceive between planning and design in landscape architecture. Weller (2006: 71) argues that landscape planning is traditionally associated with scientific instrumentality, whereas landscape design is more closely related to art. Weller (2006: 71) argues that the failure to bridge this gap in landscape architecture is caused by 'the absence of infrastructure'. With the term 'absence', Weller refers to the popular understanding that landscape architectural work deals with places lacking infrastructure or is a process of making infrastructure invisible. In other words, the common understanding of landscape architectural practice is that it beautifies places by covering un-artful infrastructure. Weller (2006: 83) further suggests that 'planning's pretences to the whole' and 'design's preoccupation with parts' should come together 'in a more finely tuned and instrumental landscape architecture'. Weller (2006: 83) assumes that convergence to be the 'ecological art of instrumentality' and suggests that it is 'in landscape architecture's favour'. It could be argued that the perceived gap between planning and design in landscape architecture is related to the lack of skills in interpreting ecological infrastructure in the art of landscape and through creative means, as argued by Corner (2006).

These arguments suggest that a common thread among the debates about the capacity of landscape architecture to address the challenge posed by an urbanising world is an emphasis on ecology and the integration of previously separate skills, knowledge and fields. These debates can be traced first to Smith's (1994) claim that new landscape forms might be inspired by a form of contemporary culture that welcomes both the vernacular and high culture. That claim is similar to Shannon's (2006) argument that landscape architects are currently positioned between the heritage of the low culture of agrarian peasantry and the high culture of urban design.

It could be argued that hybrid forms of practice and the skills of integration require landscape architects to collaborate and negotiate with the various professions in the built environment for the purpose of transforming land dominated by nature into a complex urban form. It could be even argued that landscape architects may act as consensus builders in the practice of urbanising landscape. In a recent article, *Changing Places: Where History and Nature Collide*, Arvidson and Martin (2007: 48) argue that the different agencies and stakeholder groups that landscape architects work with and for often claim authenticity of place based on their own rights. Therefore, landscape architects are often situated at the centre of a controversial redesign. In their argument for seeking a common ground between constituencies protecting historic and ecological values, Arvidson and Martin (2007: 57) claim that landscape architects act as 'facilitators' and use their multidisciplinary skills in ecology, historical research and public participation to build a common language among diverse stakeholders when discussing historic character and ecologic processes.

These discussions suggest that the role of landscape architects as consensus builders facilitates their work internationally. In other words, as a result of shared global forces of change—urbanisation, landscape architects are prepared to conduct work across cultures, to positively influence land development and to deal with degraded nature and diversified cultures. In addition to the unique approach of landscape architecture to place, landscape architects' multidisciplinary skills and their expertise in ecology, historical research and public participation (i.e., their skills in integration and collaboration) suggest the potentially prominent position of landscape architecture in the international market of planning and design.

2.3 Challenges to International Design Practice

As a dominant field of international design practice including landscape architecture, Western design practice in Asia's urbanisation appears problematic. In his discussion of the need to understand local context, Cooper (2003: 12) argues,

> In design and development circles, these arguments [about the appropriateness of cross-cultural exchange] are often couched in vocabularies of understanding and responding to "context" in order, for instance, to protect local identities and cultural heritages…a key issue in this debate is the (mis)use of information and technologies when they are moved from one culture to another, typically from a more developed country to a less developed one—more rarely vice versa.

Cooper's argument suggests that place delivery in international design practice is challenged by greater cultural difference and geographical distance between international practitioners and the socio-cultural context of a specific place. Further, the exchange of knowledge across cultures through planning and design practice has multiple dimensions.

2.3.1 Cultural Difference and Geographical Distance

In international design practice, the cultural difference and geographical distance between designers from elsewhere and a specific place are greater than the distances for design practice in practitioners' own countries. This is particularly the case of Western design practice in Asia's urbanisation. In this context, international practitioners are often positioned as relatively ignorant of the significant but hidden aspects of place, as these aspects comprise local community's life-experience and the consequently attached meanings. In his book, *Place and Placelessness*, Relph (1976: preface), the Canadian geographer and professor in planning science, pointed out that 'place and sense of place…are inextricably bound with all the hopes, frustrations and confusions of life'. Tuan (1977), the American-Chinese geographer who initiated the shift of geographical study into cultural-geography with his influential book *Space and place: the perspective of experience*, argued that a sense of place is associated with ease, comfort and familiarity. Hough (1990: 5) also claims that 'understanding places begins with feelings', and memories are often crowded with disconnected smells, sounds and sights when images of places remembered arise.

Relph (1976: 142) linked the abovementioned aspects to 'insideness', which is related to individuals, and often communities, residing in places. In contrast, this insideness immediately casts international designers as outsiders whose understandings of places often begin with rationalities. In international design practice, the detached position of international practitioners is obvious, as they may have never lived in the place that they change through their planning and design work, nor have they belonged to the social cultural context accommodating the place.

The inevitably detached position of international practitioners thus calls for more sensitive and open-minded approaches in order to know place deeply and sufficiently. Relph (1976: 142) argued that the experience of insideness of a place has four levels. The first level is authentic and unselfconscious, relating to individual, personal experience; it is also intersubjective that other people can sympathise. Relph indicated that this level of insideness is the essence of a sense of place. The second level is also authentic and unselfconscious, but relates to culture and community, rather than individual values. It is as deep as the first level, but is more associated with sacred and secular experience. The third level is shallower. It is authentic but self-conscious. Relph (1976: 142) related this third level experience to the experience of 'a sensitive and open-minded outsider seeking to grasp places for what they are to those who dwell in them and for what they mean to him[self]'. The last is the superficial level, simply being in a place and disregarding the qualities and significance of it. Relph's argument for the third level of the experience of insideness suggests the possibility that even designers from elsewhere and those practicing internationally can respond to the authentic sense of place. By contrast, their relevance to the fourth level of experience may result in them introducing placelessness through their planning and design.

The third level of experience also suggests that a more sensitive and open-minded approach is a more personal approach. However, in present-day planning and design practice, whether it is local or international, the widely adopted attitude towards place is rationality, and the consequent approach is impersonal. In Relph's (1976: 49) words, such rationalities shows the 'objective outsideness' of designers and planners, suggesting their detached position from a place. Relph (1976: 43) argues that the authentic sense of place is created through a continuous process of using, living-in and experiencing a place. Relph (1976: 49-52) further criticised the detached position of planners from a place as 'a deliberately adopted intellectual attitude', with the purpose to 'explain in a scientific manner the spatial organisation of places'. These approaches contradict a sense of place that is ‘in its essence both prescientific and intersubjective’ (1976: 145).

Further, Relph argued that, guided by objective outsideness, planners often deploy technique-dominated approaches and ignore the insideness of place. Relph saw these approaches as inevitable, since ‘a major aim of planning is to overcome spatial incongruities and inefficiencies’ (1976: 87). He illustrated this with examples, saying, ‘this is indeed technology-dominated planning, divorced from places as we know and experience them in our everyday lives, and quite casually ignoring or obliterating them’ (1976: 89). Relph (1976: 88) indicated that the use of quantitative techniques, the manipulation of data generated from averages of economic men, and dispassionate and impersonal approaches to both place and people are all technique-dominated approaches, and underpinned by the pursuit of efficiency. Similarly, Tuan (1977: 7) pointed out that ‘professional planners, with their urgent need to act, move too quickly to models and inventories…the rich experiential data on which these abstractions depend are easily forgotten’. This discussion suggests the tendency to adopt an impersonal approach and the attitude of objective outsideness in international design practice because a sensitive and more personal approach is likely to be hindered by socio-cultural difference, limited time availability, and geographical distance.

Moreover, a sensitive and more personal approach to international design practice actually works against the factors promoting its expansion: the increasingly rapid transportation and technological development. In Cantor’s (1997) review of contemporary trends in landscape architecture, rapid transportation and advanced technology were viewed as promoting the trend towards international practice in landscape architecture. Relph (1976: 90–92) had previously likened mass communication to rapid transportation that transmits products, people and ideas, which in turn, he argued all lead to the transmission of placelessness.

Relph (1976: 90) indicated that rapidity itself discourages social contact and reduces the direct involvement of outsiders with insiders. With the reduced need for ‘face-to-face contact’, design professionals are more likely to ‘treat problems as widespread and general rather than local and specific, and hence to propose general solutions, according to place-free dictates of current social science and planning’ (Relph 1976: 92). Moreover, mass communication encourages and transmits standardised tastes and fashion, which ‘are formulated by manufacturers, governments and design professionals’. Even when Relph (1976: 92) wrote his account, the work of planning and design was already regarded by him as producing placelessness and standardised street patterns in suburbs and international styles of architecture. In Relph’s view, contemporary economic systems embrace mass communications, mass culture, big business and powerful central authorities. Therefore, placelessness is inevitable. These factors, which Relph argued against in the 1970s, still appear relevant in present day. First, mass communications and mass culture parallel the new form of communication, the internet. Second, the demand for and practice of international design practice often involve big business and powerful central authorities, such as the China Central Television Headquarters that was designed by the Dutch architect Rem Koolhaas.

The contrast between the insideness of place experience and the objective outsideness experienced by international designers is also underpinned by the separation between those who belong to a place and those who design, use and invest in places. According to Relph (1976: 124), this separation is again due to technological development, which separates people from nature and changes the way people create and experience the landscape. Further, Relph (1976: 124–125) argued that a combination of the separation of man and nature as well as changes in society and economy are relevant to the loss of place. Relph's view is manifest in the increasing complexity and scale of present-day places. This increasing complexity becomes a major reason that place creation increasingly relies on design professionals with specific knowledge and skills. In the present day, places are increasingly equipped with the advanced technology needed to accommodate modern lives and functions. Under these conditions, design professionals from other places, states, countries, and even from other cultures are often invited to contribute their skills to a place. In short, the distance between those who re-form places and those who live there or use the places is increasing. Apparently, such a distance is manifest in international design practice.

Whilst it could be argued that approaches can be and should be adopted in international design practice that achieve Relph's third level of experience of insideness, Relph retained a pessimistic view of the capability of planning and design to address placelessness. Relph (1976: 146) claimed:

> A method developed by Christopher Alexander, …and approaches like Gordon Cullen's analysis…But these, and almost all the other procedures of environmental design are either too formal or too rigidly prescriptive, or they treat experience and meaning only as other variables capable of manipulation.

Relph's discussion of various approaches to knowing places can be taken as a measure, determining whether placelessness is amplified or reduced by practical means in international design practice. He concluded that the possibilities for planners and designers exist not in creating rootedness but in cultivating conditions for 'a lived-[in-]world of both everyday and exceptional experience' or for rootedness and care for places to develop (Relph 1976: 146). He provided, however, no single way to achieve these.

In recent decades, the characteristics of places have been regarded as being continuously re-formed as a result of global interconnectedness. The activities of forming and re-forming places as part of the urbanisation process are already dominated by those in the design professions, rather than being parts of a community's everyday life. Because professionals work across places, regions, nations and cultures, non-local elements are also introduced and remade. The potentially positive role of outsider designers, including those working internationally, is increasingly assumed by clients in the newly interconnected world. Apparently, it is based upon the belief that the reclaiming of the identity and character of present-day places is now recognised as occurring through their connections to the outside world. Massey and Jess (1995: 1) assert:

> …in the midst of this global connectedness, place and culture are restructured; on the one hand, previous coherences are being disrupted, old notions of the local place are being

interrupted by new connections with a world beyond; on the other, new claims to the – usually exclusive – character of place, and who belongs there, are being made.

Massey and Jess (1995: 1) explain that the interconnections of contemporary global economics, culture and ecology now disturb the internally generated uniqueness of place, suggesting that the term 'place' is being reconceptualised.

The above discussion reveals a shift in the idea of the characteristics of place in recent decades. This shift promotes the recognition that place is being continuously re-formed in global connectedness. The shift also promotes concerns for a constructive approach to re-forming place in this global connectedness. Such a concern is apparent in Seddon's inquiries into the cultural component of sense of place in the late 1990s and in his doubts about the need to design only in local context. Referring to the design of the Bauhaus and International styles, Seddon (1997: 140–141) even claims that 'some of their best and dramatic effects are achieved by contrast, 'unrelieved harmony can cloy', and 'a little discord can heighten our sense of harmony, just as the international can dramatise the local'. A positive attitude towards the infusion of non-local elements into local context is clearly meant.

In such a context, international practitioners and local stakeholders are to be positioned. Corner (1999: 12) argues that the insider perspective allows 'a deeper, socially informed, material sense of place and being' and 'grounds a project in the social practices and physical conditions of a locality'. The outsider perspective, however, suggests 'a broader range of possibilities to be invoked beyond those of the known and the everyday' and 'brings a new and broader range of ideas to bear upon the site'. Such discussions suggest that first, characteristics of place remain central in the discourse of place, although place is increasingly acknowledged as beyond being bounded or enclosed geographically or culturally. Second, cultural transformation of place is now foreseen as inevitable. Similarly, in planning and design discourse, although informed by new recognition of the formation of present-day places, the argument remains centred on ways to sustain their characteristics. More importantly, these arguments suggest that although planners and designers are increasingly from outside, the renewal of places should be based upon deeper and sufficient knowledge of a particular place, especially its history and the non-material components. They also suggest that the values that both insiders and outsiders contribute to place is now situated in global dynamics, and that approaches that bridge the perspectives of insiders (those who belong to a place, society and culture) and outsiders (international practitioners) have become crucial to the positive development of places and the continuity of their characteristics.

2.3.2 Gaining Local Understanding in Practice

In the context of international design practice, the process of gaining local understanding is more complicated than design practices within the practitioners' own societies and cultures. The socio-cultural context, the expectations of clients and users, and the procedures of project delivery can be very different from the practitioners' own. Further, in order to gain the above understandings, international practitioners need to engage with local stakeholders. The process of exchanging knowledge through the engagement with diverse local stakeholders, however, can be slower and more complicated than gaining knowledge from publications and documents.

In terms of understanding context in practice, the values of a place are relatively less recognisable to practitioners than to researchers, and it is a greater problem when practitioners are international. This limiting factor is raised in the book, *Buildings, culture and environment: informing local and global practices*, where Lorch and Cole (2003: viii) point out that:

> [A] limiting factor [to practitioners] was the inability to articulate, understand and account for the local user expectations, social and cultural values as well as ways of life.
>
> Although this is immediately obvious (especially to historians, social scientists and others), it has not been operationalised by many stakeholders involved in the built environment–on both the supply and demand sides.

In their argument, the term 'supply side' refers to the construction industry that includes the research community and policy makers (Lorch and Cole 2003: vii) as well as builders, contractors, architects, engineers and other members of the design team (Seaden 2003: 131). The term 'demand side' refers to clients, customers, purchasers and users who form the demand side of construction (2003: 130).

This limiting factor may further lead to the overlooking of hidden aspects of place in international design practice. Cole and Lorch (2003: 4) argue that the understanding of a local context needs to cover a wide range of aspects including 'values, symbols, meanings and understandings, climate, resources and history with their convergence as a way of life–the practical ways in which things are performed and achieved'. In the context of local design practice, designers know more about the context of a place; have more opportunities to engage with local community directly, and often read widely around the topic of the site and explore related issues. However, in another language and culture, these are difficult to achieve. This suggests that in the context of international design practice, opportunity for gaining sufficient context-based knowledge of places is less likely.

To further explain this limiting factor, Cole and Lorch (2003: 2) compare the difference between knowledge exchange and information transfer, arguing that knowledge is not as easily transferable as information because knowledge is attached to an individual or group and is context dependent. '[K]nowledge is a deeper understanding of a subject', and it 'entails capabilities of assessment to form judgement, interpretation and understanding' (2003: 2). Information 'is simply data which can be stored and distributed' and is therefore not able to provide deeper understanding by itself; it thus limits designers' capabilities for assessment (2003: 2). From this viewpoint, Cole and Lorch suggest that, in cross-cultural transfer of knowledge and information, information is easily stored and distributed across cultures through the 're-interpretation of texts in foreign context, international conferences and more recently via information and communications technologies' (2003: 3). In contrast, the 'absorption of relocated information into knowledge could be slow, evolving and adapting to incumbent social or physical surroundings' (2003: 3). Likewise, Kohler (2003: 83) points out that, while the abstract notions of culture, sustainability, lifestyle and environment are used almost universally, they should refer to specific context.

The above discussion suggests that the process of international design practice involves knowledge based at least upon two contexts: the foreign and the host. Principles and guidelines for planning and design and practical ways to practice have to be referred to in each context. Therefore, extra work and a longer time allowance is necessary for relocation. While design scholars argue for longer time allowances for design practice internationally (Cole and Lorch 2003: 3; Bull and Boontharm 2007: 132,), they also observe the reality of limited time in international design practice. As was analysed previously, planning and design practice, whether local or international, is often driven by efficiency, suggesting the unlikelihood of allowing the sufficient absorption of relocated information into knowledge.

In contrast, information is less context-dependent and its transfer is thus quicker. The increasingly available and advanced information technology further benefits the transfer of information from one context to another. For instance, computer and internet devices are becoming more attuned to non-cultural interpretation with English buttons and transcribing programs. In the combination of technological benefits, the pursuit of efficiency, and cultural barriers, the process of international design practice is easily occupied with information transfer rather than knowledge exchange for a deeper understanding of place. This suggests the likelihood of international design practice both to act as an agent for accelerating the spread of internationalism and to adopt impersonal approaches in assessing a place.

In the context of international design practice, the informal exchange of knowledge dominates. Cole and Lorch (2003: 2) argue that the exchange of knowledge between researchers and practitioners is a 'formal exchange of explicit knowledge', and therefore, 'publication and other forms of information are important vehicles for this exchange'. In contrast, the exchange of knowledge between practitioners is less formal. Cole and Lorch (2003: 2–3) then suggest that informal exchange between practitioners is a process of gaining 'tacit knowledge' and is central to design practice, which is provided largely through the experience of the diverse members of the design team. They (2003: 2–3) also suggest that this tacit knowledge is acquired through practical activities; therefore, 'its transfer requires person-to-person contact, and individual learning and experiences'. They (2003: 3) further point out that this tacit knowledge is about the understanding of the 'how'. The gaining of tacit knowledge in the context of international design practice could occur between skilled people; however, the process is slow (2003: 3). The above discussion suggests that the interaction between local and international design practitioners is necessary in international design practice, because it helps build an understanding of the 'how' and the absorption of knowledge relocated from international practitioners' own to the local context. It also suggests that such an engagement, although direct through person-to-person contact, is also more informal. According to Michael Polanyi (1967), the physical chemist and philosopher who developed the concept of tacit knowledge, this type of informal engagement is a form of tacit knowing.

In terms of understanding expectations, international practitioners need to engage with local clients in order to build more realistic expectations. Baker (2003: 129) argues that the "the same systems of global communication are far more effective at diffusing the unsustainable expectations of the Western world", suggesting the willingness to follow Western models by non-Western clients. Such willingness is also revealed in the previous analysis of Western design practice in Asia's urbanisation, which is partially promoted by the desire of developing countries to catch up with the West and to respond to historical asymmetry. It can be argued that while globalisation promotes the prevalence of western values, such values may impact on local expectations, challenging foreign designers to distinguish between local specificities and universal ones. Baker (2003: 239) suggests that 'designers and clients can offer leadership in helping to create more realistic expectations'. Further, he (2003: 239) suggests that designers and clients need to 'adapt the vocabularies, tools, and methods from other disciplines to ensure a balance between meeting and moulding [different] expectations'. This suggests that closer interactions between designers and clients are essential for the building of consensus of various stakeholders.

In the context of international design practice, however, such leadership performed jointly by designers and clients might be challenged by different ways of thinking. With reference to Western design practice in Asian countries, it could be argued that the rooted cultural difference may have strong impacts on the exchange of knowledge between Western designers and Asian clients, influencing the building of realistic expectations. This rooted difference may even be less visible than the hidden aspect of a local context. For instance, Western sensibilities in thinking are identified as 'casual' or 'logical' (Hall and Ames 1995: 6). In contrast, the classical Chinese thinking is identified as 'aesthetic' or 'correlative', and these ways of thinking and living persist (1995: 6). From a broader perspective, the generally considered Eastern and Asian thoughts are rooted in classical Chinese thinking (Bull and Radovic et al. 2007: 221). The classical Chinese, or Asian correlative thinking in Radovic's words, may not invite Western rational objectivity (Hall and Ames 1995). Cook (2003: 365) argues that exercising human values in a foreign culture is often loaded with risk, challenging individuals and confronting institutions and corporations even more. This suggests that the collaboration between designers and local clients is more difficult in larger and more complicated international projects that involve a range of institutions and organisations. This also leads to the third dimension of local understanding—the delivery process.

Understanding the delivery process can ensure the realisation of planning and design concepts and be more important for dealing with large international projects. Looking specifically at the process of design and construction delivery, Cook (2003) argues that the delivery process in the built environment has not changed but has become much larger and more complex with globalisation. He (2003: 369) points out that the larger and more complicated delivery process requires more levels of understanding and participation as well as 'a heightening of engagement of every level of delivery'. Cook (2003: 370) suggests that old delivery methods such as the 'iterative design process' remain useful to inform new practice. This old method includes conscious practice by exercising continuity, documenting projects and critically assessing experience (2003: 370). However, he argues that a better delivery process requires the improvement of 'people-to-people skills' (2003: 370).

People-to-people skills, however, can be overlooked in the often-comprehensive structure of large international design firms, and in international projects that are often large in size and complicated in operation. It can be argued that when organisations become larger and more complex, they have to rely more on standards and techniques, and direct person-to-person engagement is thus be reduced. Michaud (2003: 269) recommends engagement with different institutions involved in international projects, saying that it may improve cross-cultural exchange of both technology and knowledge. He (2003: 269) argues that such institutional engagement can reveal social and organisational understanding of stakeholder interests and may further release various sources of 'latent' institutional, organisational, and technical innovation.

A common solution for the understanding of context, expectation and delivery process, as shown in the above arguments, is effective and full level of engagement through person-to-person interaction. This includes engagement with local practitioners to define ways to adapt the knowledge of the international practitioner into the other context, with local clients to build realistic expectations, and with local institutions to ensure a quality project delivery. In relation to Western design practice in Asian countries, the person-to-person contact is between different cultural groups. Therefore, the need for inter-personal skills and the improvement of those skills is ultimately linked to the need for cultural skills, such as ways of involving people and methods of communication.

Cultural skills are therefore essential for gaining local understanding in international design practice. From the perspective of cross-cultural design practice and explicit to urban design, including landscape architectural design, Bull and Boontharm (2007: 134) suggest that both technical and cultural skills are essential for planning or design practices situated in complex systems and characterised by difference. However, they (2007: 134) argue that in contemporary design practice, cultural skills especially need to be developed.

In their propositions about ways to facilitate cross-cultural dialogue and practice, Bull and Radovic et al. (2007: 212–220) highlight four aspects of improving cultural skills. The first aspect is communicative formats that can facilitate exchange, generate ideas, encourage reflection upon, and dissimilate the experience of practice. Symposia, workshops and studios were raised as useful communicative formats to achieve better engagement amongst various participants from different cultural and disciplinary backgrounds. These communicative formats are ways to achieve direct person-to-person contact. In addition, Bull and Radovic et al. (2007) argue that language difference is not a fundamental barrier to cross-cultural engagement, suggesting that cross-cultural knowledge exchange relies as much on other ways of engaging as on the technology of communication. The second aspect regards the conceptual open-mindedness of practitioners in cross-cultural learning. The third aspect is the use of culturally responsive methodologies that respect the ethical and philosophical traditions of the people and places that they serve. Finally, the authors proposed that practitioners should be ready to face and accept non-standard outcomes that reflect both local and global conditions.

Such propositions reveal how designers can facilitate engagement, investigate and learn from the locale and be open-minded about a more uncertain future, which is a characteristic of our global era when urban and landscape architectural practices are increasingly cross-cultural. Such thinking also suggests that in the context of international design practice, a full level of engagement may include local clients, local disciplinary experts, local communities, local design teams, and various local institutions. These local stakeholders therefore include more than the economically defined clients, customers, purchasers and users, as was argued against by Relph even in the 1970s.

2.4 Conclusion

Chapter Two identified that landscape architecture's approaches to place and practice have the primary goal to sustain the characteristics of place. This goal leads to the concern for hidden aspects of a particular place, and to the emphasis on ways to gain understanding of the underlying meaning, socio-cultural context and natural system of place. Therefore, landscape architects immerse themselves into a site or place through personal experience, carefully investigating sites, engaging with local communities, and conducting substantial research of the history of the place. In their practice for the urbanising landscape, landscape architects extend their knowledge from that of site or place to that of the landscape setting, their working territory form design to planning, and develop skills of integration and collaboration for the cultural construction and ecological recovery of place. The sustained characteristics of place should comprise both the natural and cultural, the physical and spiritual, the visible and invisible, and the long-established and newly introduced.

The great scale of urbanisation occurring in Asian countries has induced a notable field for international design practice, suggesting a potentially prominent position for landscape architecture. The inviting of Western design practice by Asian clients creates possibilities for the exchange of knowledge across cultures. This two-way dimension suggests a potential cross-fertilisation process that international design practice activates. In this context, Western design practice, as a part of the urbanisation of Asia, may provide valuable references for the world and influence working methods, design paradigms and ideologies of both Western designers and local practitioners in Asia. Landscape architects' multidisciplinary and synthetic skills, expertise in ecology, and strategies for collaborating with other professions all suggest a role for the landscape architect as consensus builder amongst diverse stakeholders of international projects. These disciplinary-specific skills can meet the complexity that international projects often present and facilitate landscape architects' acquirement of local understanding, and to positively intervene in the re-formation of place.

Gaining local understanding in international design practice not only involves acquiring knowledge of the meaning of place and its cultural context and natural systems, but also the expectations of diverse local stakeholders and the system of project operation. Such an understanding extends the scope of professional knowledge beyond the understanding of sites, landscapes and place-based community, to the understanding of diverse groups of participants and the process of project delivery, both of which emphasise personal engagement, a tacit form of knowing. The process of gaining local understanding involves information transfer and exchanges of knowledge. While the former requires technical skills, exchanges of knowledge through personal engagement require effective and sensitive methods of communication. These methods are aimed at the identification of the local context, expectations and process, in both their explicit and implicit forms.

References

Arvidson, A. R., & Martin, F. E. (2007). Changing places: Where history and nature collide. *Landscape Architecture, 97*(11), 48–57.

Armstrong, H. (1999). Landscape design in Australia. *Suzhou Institute of Urban Construction and Environmental Protection, 12*(3), 30–34.

Baker, N. V., Lorch, R., & Cole, R. (2003). Reconciling expectations. *Buildings, culture & environment: Informing local & global practices* (pp. 221–240). Oxford, Malden, MA, UK: Blackwell Publication.

Bull, C., Boontharm, D., et al. (eds.). (2007). *Cross-cultural urban design: Global or local practice?*, London.

Bull, C. J., Radovic, D., et al. (2007). Urban design for a cross-cultural future. In C. Bull, D. Boontharm, & C. Parin (Eds.), *Cross-cultural urban design: Global or local practice?* (pp. 208–233). Routledge.

Bull, C., & Boontharm, D. (2007). Reflecting on cross-cultural interactions. *Cross-cultural urban design: Global or local practice*? (pp. 129–134). Routledge.

Cantor, S. L. (1997). *Contemporary trends in landscape architecture.* New York: Wiley.

Cooper, I., Lorch, R., & Cole, R. (2003). Understanding context. *Buildings, culture & environment: Informing local & global practices* (pp. 11–17). Blackwell Publication, Malden, Oxford, UK.

Corner, J. (Ed.). (1999). *Recovering landscape: Essays in contemporary landscape architecture.* New York.

Corner. J. (2006). Terra flux. *The Landscape Urbanism Reader,* 21–33.

Cole, R., & Lorch, R. (2003). Introduction: knowledge, values and building design. *Buildings, culture and environment: Informing local and global practices* (pp. 1–8). Oxford, Malden, UK.

Cook J, Lorch R, Cole R (2003). Understanding delivery process. *Buildings, culture & environment: informing local & global practices* (pp. 359–371). Oxford, Malden, UK: Blackwell Publication.

Czerniak, J. (2006). Looking back at landscape urbanism: Speculations on site. *The Landscape Urbanism Reader*, 105–123.

Dutt, A. K. (2003). *Challenges to Asian urbanization in the 21st century.* Kluwer Academic Publishers.

Fieldhouse, K., & Harvey, S. (1992). *Landscape design: An international survey.* London : Laurence King.

Girot, C. (1999). Four trace concepts in landscape architecture. *Recovering landscape: Essays in contemporary landscape architecture* (pp. 59–67). New York: Princeton Architectural Press.

Hall, D. L., & Ames, R. T. (1995). *Anticipating China: Thinking through the narratives of Chinese and western culture.* State University of New York Press.

Hall, P. G., Pfeiffer, U. (2000). *Urban future 21: A global agenda for twenty-first century cities.* New York : E&FN Spon.

Hebbert, M., Sonne, W., Monclus, J, & Guardia, M. (2006). History builds the town: On the uses of history in twentieth-century city planning. *Culture, urbanism and planning* (pp. 3–19). Aldershot, Burlington, England: VT Ashgate.

Hough, M. (1990). *Out of place: Restoring identity to the regional landscape.* Yale University Press.

King, R. (2007). *Confronting epistemes. cross-cultural urban design: Global or local practice?* (pp. 17–26). London: Routledge.

Kohler, N., Lorch, R., & Cole, R. (2003). Cultural issues for a sustainable built environment. *Buildings, culture & environment : informing local & global practices* (pp. 81–108). Oxford, Malden, UK: Blackwell Publication.

Lorch, R., & Cole, R. (Eds.). (2003). *Buildings, culture and environment: Informing local and global practices.* Malden, MA, Blackwell Pub.

Marshall, R. (2003). *Emerging urbanity: Global urban projects in the Asia Pacific Rim.* Spon Press.

Mathur, A. (1999). Neither wilderness nor home: the Indian Maidan. In J. Corner (Ed.), *Recovering landscape: essays in contemporary landscape architecture* (pp. 205–220). New York, Princeton Architectural: Press.

Massey, D. B., & Jess, P. (Eds.). (1995). *A Place in the world?: Places, cultures and globalization.* Oxford, England; New York, Open University; Oxford University Press.

Michaud, P., Lorch, R., & Cole, R. (2003). Social and organizational understanding of stakeholder interest. *Buildings, culture & environment: informing local & global practices* (pp. 267–287). Oxford, Malden, UK: Blackwell Publication.

Missingham, G. K. (2007). Japan 10±, China 1: A first attempt at explaining the numerical discrepancy between Japanese-style gardens outside Japan and Chinese-style gardens outside China. *Landscape Research, 32*(2), 117–146.

Monclus, J., Guardia, M. eds. (2006). Culture, urbanism and planning. *Heritage, culture, and identity.* Aldershot, Burlington, England: VT Ashgate.

Neilson, L. (2000). Property council of Australia. *Asian urbanisation and the impacts on Australian cities.* Brisbane: Property Council of Australia.

Ness, G. D., Talwar, P. P., et al. (2005). *Asian urbanization in the new millennium.* Singapore: Marshall Cavendish Academic.

Paterson, G., Pettus, A. T., Edquist, H., & Bird, V. (1994). The visible landscape. *The culture of landscape architecture* (pp. 139–154). Melbourne : Vic., Edge Publishing in association with the Department of Planning Policy and Landscape RMIT.

Polanyi, M. (1967). *The tacit dimension.* London: Routledge & K. Paul.

Pomeroy, G., & Webster, G. (Eds.). (2008). *Global perspectives on urbanization.* Lanham.

Relph, E. C. (1976). *Place and placelessness.* Pion.

Rogers, E. B. (2001). *Landscape design: A cultural and architectural history.* New York: Harry N. Abrams.

Seaden, G. (2003). The role of the client in shaping the satisfactory outcome of the construction process. In R. Lorch & R. Cole (eds.), *Buildings, culture & environment: Informing local & global practices* (pp. 130–153). Oxford, UK, Malden, MA: Blackwell Pub.

Seddon, G. (1997). *Landprints: Reflections on place and landscape.* Cambridge University Press.

Shannon, K. (2006). From theory to resistance: Landscape urbanism in Europe. *The Landscape Urbanism Reader*, pp. 141–161.

Simmonds, R., & Hack, G. (2000). *Global city regions: their emerging forms.* London: E & FN Spon.

Smith, K. (1994). Relics, prosthetics and surrogate realities. *The Culture of Landscape Architecture*, 211–221.

Spens, M. (Ed.). (2007). *Landscape architecture: Site/non-site.* Chichester.

Tuan, Y.-F. (1977). *Space and place: The perspective of experience.* University of Minnesota Press.

Wall, A. (1999). Programming the urban surface. In J. Corner (Ed.), *Recovering landscape: essays in contemporary landscape architecture* (pp. 233–250). Press.

Weller, R. (2006). An art of instrumentallity: thinking through landscape urbanism. *The landscape urbanism reader* (pp. 69–85). New York: Princeton Architectural Press.

Zhu, J. (2004). Modernization: in search of Chang Yonghe and his extraordinary architecture in a historical matrix. *Architects* (建筑师) (4), 14–17.
Zhu, J. (2005). Criticality in between China and the West. *The Journal of Architecture 10*(5), 479-498.
Zhu, T. (2006). The 'Criticality' debate in the west and the architectural situation in China: Thoughts on the essay "criticality in between China and the west". *Time-Architecture* (5), 71–78.

Chapter 3
Research Design

Abstract This chapter proposes research questions that are informed by the review of literature. It also discusses the methods used in the following studies of international practice in landscape architecture. This chapter comprises five sections. Section 3.1 explains the research questions designed to test the research hypothesis of this thesis, and discusses the need for multi-faceted data collection to address these questions. Section 3.2 explains the selection of methods, including a review of key texts, survey and case study methods. Section 3.3 explains the purpose and scope for the review of key texts along with the survey and case study. Section 3.4 explains the application of survey methods in this research, including a discussion of the design of the survey questionnaire, the procedures for data collection, and the analytical approaches. Section 3.5 explains the application of the case study method in this research, including case selection, sources of data, procedures for data collection and analytical approaches.

Keywords Questionnaire survey · Interviews · Case study · Projects in China · Design documents

3.1 Research Hypothesis and Questions

Chapter 1 introduced the aim of this research, which was to explore the characteristics of international practice in landscape architecture. Under this aim, approaches to knowing and re-forming places that have been theorised in landscape architecture and discussed in relation to international design practice were investigated in Chap. 2. The review of theory brings out three propositions of further study. First, landscape architecture's approaches to place and practice have the primary goal to sustain the characteristics of place. Second, landscape architecture has a prominent position in the international market of planning and design practice. Third, landscape architects' role of consensus builders supports their delivery of complex projects that Asia's urbanisation presents. Based upon these propositions, the research further proposes a hypothesis that directs the followed study of international practice in landscape architecture:

Y. Zhang, *Place Making in International Practice of Landscape Architecture*, Advanced Topics in Science and Technology in China 64,
https://doi.org/10.1007/978-981-16-2442-1_3

> Landscape architects use their disciplinary-specific skills to gain local understanding and act as consensus builders across cultures and multidisciplinary teams to sustain the characteristics of place during urban development.

To test the above hypothesis, the thesis proposes two research questions, specifically for Australia and China:

1. What are the characteristics of Australian landscape architectural practice internationally and specifically, in China?
2. How do Australian landscape architects engage with the characteristics of place in China when they practice there?

The first question concerns the recent phenomenon of Australian landscape architectural practice internationally, with a focus on China. The second question concerns the process of the practice of Australian landscape architects in China, and the consequent outcomes. To answer these research questions, a number of methods were selected and are discussed below.

3.2 Selection of Methods

The research questions suggest the need for multi-faceted data, such as that relating to the types of firms providing landscape architectural services internationally, the types of projects put out to these firms, local expectations of international landscape architectural practice, the procedures for providing these services, and the consequent outcomes. Therefore, multiple methods are needed. These methods included the review of key texts that were relevant to Australia's landscape architectural practice internationally, and China as a site of demand for international planning and design services, including landscape architecture. These methods also included survey methods and case study methods because surveys are often adopted to answer the question of 'what' (Yin 2003, p. 5) (the first research question), whereas case studies are often used to answer the question of 'how' (Yin 2003, p. 1) (the second research question). Therefore, the three major methods adopted to conduct the following study are as follows:

1. The review of key texts
2. Survey methods
3. Case study methods.

The following sections explain respectively the purpose and scope of the review of key texts, the survey methods and application, and the case study methods and application.

3.3 The Review of Key Texts

The review of key texts, along with the application of survey methods, was for collecting data in two areas. The first area of data was the background of Australia's supply of landscape architectural services internationally and in China. Therefore, the review focused on literature on Australia's geographical and cultural positions in Asia, Australia's relations with China, and relevant statistics of Australia's exports of services. The review of this part is presented in Chap. 4.

The second area of data was about China as a site of demand for international planning and design services, including landscape architecture. Therefore, the review focused on literature on China's reform and urban development in recent decades, the infusion of international planning and design practice, and the status of the local profession of landscape architecture. The review of this part is presented in Chap. 5.

The review of key texts, along with the application of case study methods, was for collecting data relevant to the selected case project and is in two aspects. The first aspect was the regional and local characteristics of the selected area in China where Australian landscape architects provided services. The review focused on literature on the regional context and the history of the area. The analysis of this part is presented in Chap. 6. The other area of data was the context of the case project and the project outcomes. Therefore, the review focused on key texts relevant to the initiation of the case project and the comments on the outcomes. The analysis of these data is presented in Chaps. 6, 7 and 8. A further explanation of the sources of this aspect of data is in Sect. 3.5: Case study methods and application.

3.4 Survey Methods and Application

The first research question, on *what are the characteristics of Australian landscape architectural practice internationally and specifically, in China*, required timely information on how many and in what ways Australian landscape architectural businesses participated in the international and the Chinese market of planning and design. The size of firms and types of international projects and services were of direct relevance. Because survey is a method of collecting information directly from people (Fink and Kosecoff 1998, p. 1) and participants can provide descriptions of their attitudes, values, and background characteristics (1998, p. 3), this method was useful for collecting information directly from those Australian landscape architects who had practiced or were currently practicing internationally and in China.

The survey group was those had registered their businesses with the Australian Institute of Landscape Architects (AILA), the national professional organisation. The pre-designated respondents to the survey were directors and individual practitioners of those registered businesses. A list of registered businesses names in the eight states and territories of Australia was sourced from the official website of the AILA (http://www.aila.org.au/). The eight states include the Australian Capital Territory, New

South Wales, Northern Territory, Queensland, South Australia, Tasmania, Victoria, and West Australia. The number of the registered businesses was 272 in the survey period, which started in August 2007.

The method for collecting survey data was through mail-out questionnaire because mail-out questionnaires had advantages over web-based, telephone and in-person interviews (Rea and Parker 2005, p. 8). According to Rea and Parker (2005, pp. 8–11), the mail-out questionnaire survey has eight advantages:

1. Cost savings from lack of recruitment and travel;
2. Convenience for respondents;
3. Ample time for response;
4. Legitimacy and credibility of mail-out questionnaires;
5. Anonymity for participants;
6. Reduced interviewer-induced bias;
7. Questionnaire can be longer and more complex than verbal questions;
8. The possibility to use other visual aides to support questions, such as tables and figures.

The above advantages were important for this research because it needed to collect concrete and comprehensive data with a limited budget and time allowance. The disadvantages of such a method include comparatively long response periods, potentially low response rates, lack of interviewer involvement, and potentially incomplete responses to open-ended questions. The questionnaire was therefore designed to be answered in approximately fifteen minutes and was tested before mailing out. A two-week period for response was then set for the questionnaire. After the receipt of responses and primary analysis, follow-up emails were sent for further clarification if needed. In addition, interviews were designed as part of the case study methods. This reduced the disadvantage of a survey that lacks interviewer involvement. The method of interview is explained in the next section. On 29 August 2007, printed questionnaires were posted to the 272 business addresses registered with the AILA. A full list of questions is included in the Appendix.

The design of the questionnaire was aimed at collecting three types of data: descriptive, behavioural and attitudinal information (Rea and Parker 2005, p. 6). Based upon these three types of information, eight questions comprising three themes of inquiry were designed. The first theme comprised two questions and explored the characteristics of firms conducting landscape architectural work internationally. It established the size of firms practicing internationally and the income of their international work. The second theme of inquiry comprised five questions and explored the characteristics of the work provided by Australian landscape architectural businesses for China. It established the types of projects, the types of services provided, the methods of engagement, the sources of work, and whether these Australian firms had actively sought work in China. Data collected through the above two themes was quantitative. The third theme of inquiry, however, explored the qualitative aspect of Australian landscape architectural practices in China. Through an open-ended question, it inquired whether Australian landscape architects faced challenges posed by providing services for China and the strategies used to meet these challenges.

To build a more focused look into Australia's provision of landscape architectural service for China, the questionnaire focused on the period from 1995 to 2007. The selection of this most recent period enabled the collection of timely information on the provision of landscape architectural services from Australia in China. It also enabled the relation to China as a site of demand for international planning and design services during this particular period, when China was in rapid urbanisation and modernisation.

Of the 272 mailed questionnaires, 104 responses were received. A further thirteen were returned unopened, suggesting a change of business addresses during the period when the survey was conducted. Therefore, an approximately 40% response rate was achieved in this survey. Fowler (2002: 40) argues that 'the response rate is a basic parameter of evaluating the data collection effort,' yet 'there is no agreed-upon standard for a minimum acceptable response rate'. Fowler (2002, p. 42) further indicates that 'for most mail surveys, people who have a particular interest in the subject matter or the research itself are more likely to return questionnaires than those who are less interested'. One example is that younger people are less likely to return questionnaires studying health care experiences (Fowler 2002, p. 42). Therefore, 'the respondent's interest may play a bigger role in the response to mail surveys' (Fowler 2002, p. 44). As the subject of this survey was Australian landscape architectural practice internationally and specifically, in China, the 40% response rate itself could then explain, to some extent, Australian practitioners' interest in providing landscape architectural services internationally and in China. In addition, the 40% response rate was close to the suggestion of Rea and Parker (2005, p. 11) that a 50% response rate was satisfactory for a mail-out survey. However, Rea and Parker's estimate did not consider participants' interest in the research. According to the argument of Fowler, the 40% response rate of this survey was deemed satisfactory for the purposes of analysis and the reporting of data.

The method used for analysing the survey data was the common 'descriptive statistics' (Fink and Kosecoff 1998, p. 59), which are also the basis of more advanced methods (Fink and Kosecoff 1998, p. 60). The analytic method of descriptive statistics includes counts (numbers or frequencies), proportions (percentage), measures of central tendency (the mean, median and mode) and measures of variations (range and standard deviation). These techniques are useful to process the data so that it can be interpreted in the form of numbers, proportions and scales. Computer software, Microsoft Excel, was used to work out frequencies and trends based upon the survey data. This formed the quantitative analysis and interpretation of the survey data.

The survey data collected through the open-ended question was about Australian practitioners' understanding of strategies to meet challenges presented by providing service for projects in China. Such data was non-numeric but qualitative. Strauss and Corbin (1998, pp. 10–11) suggest that research that 'produces findings not arrived at by statistic procedures or other means of quantification' was qualitative research. Thus, the procedure for interpreting and organising the data collected through the open-ended question adopted a qualitative method of analysis. This was by 'conceptualising and reducing data, elaborating categories… and relating [them] through a series of prepositional statements' (Strauss and Corbin 1998, p. 12). The data were analysed in relation to the propositions of this research that ground its hypothesis. This is explained further in the following section. The analysis of all of the above quantitative and qualitative survey data is interpreted in Chap. 5.

3.5 Case Study Methods and Application

The second research question, on how Australian landscape architects engage with the characteristics of place in China when they practice there, was about the process of the practice. This question can best be informed by investigating the real life context of Australian landscape architects providing service for a project in China. The context of the project and the expectation from international design teams and landscape architects, the composition of these teams, the ways that explicit and implicit knowledge of place were obtained, the methods for engaging with relevant local stakeholders, the interpretation of local characteristics in planning and design, and the consequent project outcomes, were all relevant. To explore such issues, however, multiple types of evidence are required.

The method of case study is useful for collecting and analysing the required evidence mentioned above. Yin (2003, p. 1) argues that 'case studies are the preferred strategy when how or why questions are being posed, when the investigator has little control over events, and when the focus is on a contemporary phenomena with some real life context'. Compared with other major research strategies for social science, such as experiments, surveys, archival analyses and histories that deal with evidence in documents and artefacts, the unique strength of a case study is its ability to deal with a full variety of evidence, including those acquired from interviews and observations (Yin 2003, p. 8). Francis (2001, p. 15), a professor and past chair of landscape architecture at the University of California, argues that 'case studies have a long and well established history in landscape architecture' and 'often serve to make concrete what are often generalizations or purely anecdotal information about projects and processes'. The study of how Australian landscape architects practiced in China and engaged with the characteristics of place was therefore well suited to the case study method of data collection and analysis.

The case study adopted single case study based on the work of Australian landscape architects for an urban development project in China. Yin (2003, p. 40) argues that one rationale for a single case study is when it presents 'a critical case in testing a well-formulated theory'. Yin (2003, p. 40) points out:

> The theory has specified a clear set of propositions, as well as the circumstances, within the propositions that are believed to be true. To confirm, challenge or extend the theory, a single case may meet all of the conditions to test the theory. The single case can then be used to determine whether a theory's propositions are correct or whether some alternative set of explanations might be more relevant. In this manner…the single case can represent a significant contribution to knowledge and theory building.

As proposed in the first section, the thesis has a research hypothesis. The hypothesis is grounded in a set of propositions that have been identified from theories of

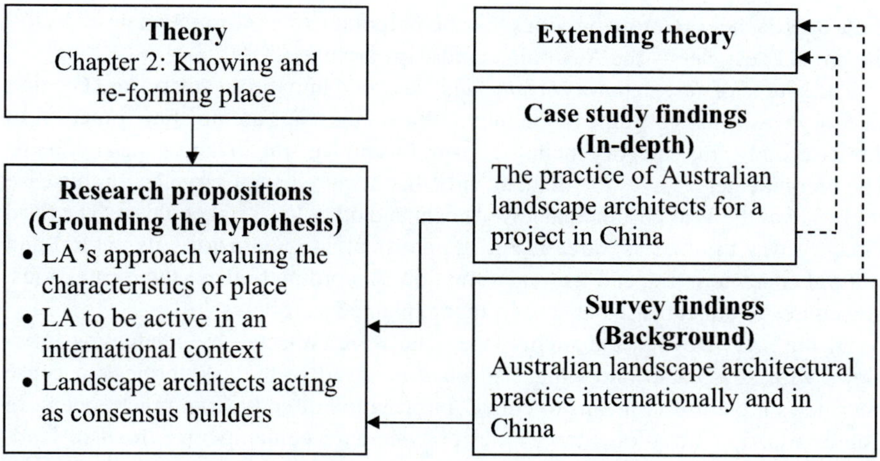

Fig. 3.1 Research design for extending theories. *Source* Adapted from *Case Study Research: Design and Methods* (Yin 2003, p. 32)

place, landscape architecture and international design practice. Therefore, according to Yin (2003, p. 45), the single case study of this research can represent 'a critical test of the existing theory' and expand current knowledge. Figure 3.1 explains the approach of this research to contribute to the current knowledge of landscape architecture and international design practice.

According to Francis (2001, p. 18), using a case study method in landscape architecture, however, has three major limitations. One is the length of time needed on-site. The second is the inclination of project designers, owners and managers not to provide information about the problems in their projects. The third is that assessment of some landscape architectural projects may be better conducted after a few years, even a decade or more. Ways to overcome these limitations were considered in this research and addressed though repeated on-site investigations, conversations with park users and local supporters of the project, and a review of publicly assessable information reporting the project.

The selected case project was the Planning and Construction of the Li Lake Area in Wuxi, China, which involved two international design teams and a range of Australian landscape architects for each team. The case study focused on the four-year period from the beginning of 2002 to the end of 2005, when the two international design teams provided consultancy services of planning, landscape design, landscape design development and construction supervision. Three trips were made to the project site, one in April 2007 and the other two in November and December 2008 respectively. Therefore, the on-site studies were conducted years after the completion of the project, by which time the conditions of public use were observable and public opinion became available. During the last trip, conversations were also had with local practitioners and governmental clients. The information collected through these approaches therefore provides the counterpoint to the information provided by the project designers—the Australian landscape architects.

The source of case data was diverse and classified into three categorises. The first category was the supply side of the international landscape architectural practice. In this case study, this category included Australia and Australian landscape architects. The second category was the demand side that acquired such services. In this case study, it was the Wuxi Municipal government and other local stakeholders. The third category was the Li Lake area itself, the physical site, its history, its natural and cultural characteristics, and its transformation. According to these three categories, procedures to collect case data are further explained as follows.

Of the first category, the supply side, data were collected by conducting interviews with key practitioners that had been involved in the case project, because interviews are a useful means to collect information directly from practitioners. In Cantor's (1997) study of international practice as a contemporary trend in landscape architecture, he also used interviews as sources of information on the projects and to learn practitioners' approaches to design. In this case study, three interviews were conducted. The first was the interview in late September 2007 in Melbourne, Australia, with a director of the Melbourne office of the Australian firm Tract Consultants. The second was the interview in early October 2007 in Brisbane, Australia, with a director of the Brisbane office of the international firm EDAW.[1] The third was the interview in early November 2007 in Shanghai, China, with the Australian landscape architect who acted as construction supervisor of the case project and worked for the EDAW Shanghai office.

The interview questions were pre-provided to interviewees and focused on their understanding of local natural and cultural characteristics and their approaches to gaining this understanding. The interviews were semi-structured and each took approximately one hour. Upon the consent of the interviewees, all of the three interview conversations were recorded in transcripts. Through the interviews, key design documents and records of the process were acquired. Data collected in this area is listed as follows and the interview questions are provided in the appendix.

The interview in Melbourne:

1. A semi-structured conversation based upon the pre-provided interview questions.
2. The document titled *Concept Planning Design of Wuli* [Li] *Lake Area of Wuxi City*, dated May 2002.
3. A package titled *Double Bay, Tai Hu* [Lake Tai]*: Analysis, Concept Options, and Preferred Structure Plan*, dated June 2003 (another project located near Li Lake area).

The interview in Brisbane:

[1] Since 2008, EDAW has been a part of the global provider of professional technical and management support services, AECOM, which is formed from many of the world's finest engineering, design, environmental and planning companies (Retrieved Jul 08, 2010, from http://www.aecom.com/About/The+AECOM+Brand).

1. A semi-structured conversation based upon the pre-provided interview questions.
2. The document titled *New Lake* [Li Lake] *City, Wuxi, China, Master Plan Presentation*, date not specified.
3. A document titled *Landscape Design Development Package: Wuxi Lakefront—stage two*, dated July 2003.
4. A document titled *Landscape Design Development Package: Fishing Dragon Island*, dated November 2003.
5. A range of e-news articles issued by EDAW reporting the progress and success of work.
6. International and Australian awards received for the completed waterfront.

The interview in Shanghai:

1. A written reply to the pre-provided interview questions.
2. A semi-structured conversation.
3. Files of the application for awards for the completed waterfront, containing key points of success, photos with illustrations of the completed waterfront, and project costs, timeline and key contacts of the EDAW team and local stakeholders.

In addition, two packages of design documents for the case project were collected through the Internet. These design documents were then confirmed as the work of the EDAW Brisbane office by the interviewed director. These documents included:

1. The package titled *Wuxi New* [Li] *Lake City, Master Plan Report*, dated August 2002.
2. The package titled *Five Mile* [Li] *Lake Landscape Design*, dated December 2002.

Moreover, websites of the Australian firm Tract Consultants, the international firm EDAW (later AECOM), and those of other professional organisations in Australia were used to collect relevant information, including the following:

1. The AILA Queensland State Awards 2005 commemorative book.
2. The 2003 Australian planning awards received for the planning of the Li Lake area by Tract Consultants.

For the second category, the demand side, data were collected through personal communication with the Wuxi Municipal Committee of Planning and Construction of the Li Lake Area (WMCPC), because formal interviews were not acceptable by governmental officials in the local context. Personal communication also included that with the director of the local design team that provided support in the construction stage. Conversations with the members of the WMCPC and the director were conducted in their offices and recorded as notes afterwards to assist with analysis. Three books containing information on the case project were kindly provided by the WMCPC. The first book was published by the society of *Urban Science Research of Wuxi*, and issued for internal circulation within local government institutions. The second book was a pictorial record of the Li Lake area in the past and in 2007. It had English interpretations and was commercially available. The third book was a record of the progress of the case project by the WMCPC. Data were also collected through the website of the government of Wuxi. The conversation with the WMCPC, the three publications and the government website provided information of the following aspects:

1. Reasons for the development
2. Structure of the case project
3. Reasons for using international consulting teams
4. An official reply to the planning schemes provided by the international consulting teams
5. Key events and phase outcomes of the case project
6. International and Chinese national awards received for the built waterfront as a part of the case project
7. Opinions of the WMCPC and local supporters of the international participation, the project outcomes, and the sustained local characteristics.

For the third category, the Li Lake area as it was before and after the development, data were collected through a review of key texts and the observation of the completed Li Lake waterfront. The observation included two phases, the first in late April 2006, and the second in mid November 2007. Both were seasons ideal for outdoor activities and therefore ideal for observing how the designed and constructed waterfront was used. Photographs were taken during the two trips. The observation focused on places popular for park users and activities taking place there. Conversations were also conducted with waterfront users, if they were willing, and recorded as notes afterwards. The conversation focused on Wuxi residents and their opinions of the renovated waterfront, centring on two questions: their favourite places and aspects of the waterfront they thought that are better than before. These notes were subsequently used to assist with analysis. The nearby traditional Li Garden and the nearby Double Bay area were also visited and photographed. Key aspects of data collected from the third source included:

1. The regional context of the Li Lake area
2. The local history of the Li Lake area
3. The present state of the Li Lake waterfront: the physical landscape and activities of park users taking place there.

The analysis of the above data took a qualitative approach and was done by building three units of analysis. Yin (2003, p. 28) argues that any research design should have five components, including the research questions, its propositions, its units of analysis, the logic linking the data to the propositions, and the criteria for interpreting the findings. The former three are concerned with the types of data, while the latter two with analytical approaches. Therefore, the analysis of the case data started with the building of units of analysis. These units are explained as follows.

This research is concerned with ways international landscape architectural practice can engage with the characteristics of a place. The processes of international practitioners studying the place, engaging with local stakeholders, and identifying and interpreting its characteristics in planning and design were all central to the case study. Therefore, the procedure and key activities of practice, as manifested in the participation of Australian landscape architects in the case project, was a major unit of analysis. The analysis of this unit, however, needed to be supported by the analysis of local characteristics that were recognised, the local aims for the development, the governmental expectations from international landscape architectural firms, and the outcomes of the work. These supporting analyses led to another two sets of units: the context of the area and the proposed development, and the outcomes presented along with the project process undertaken by the involved Australian landscape architects.

Therefore, the collected case data were categorised into three units of analysis. The following list of units of analysis shows that each unit contains data from the three sources:

A. The context of the Li Lake area and the proposed development.
B. The project process applied by the involved Australian landscape architects.
C. The characteristics of the Li Lake area interpreted in the landscape architectural work phase outcomes.

The logic linking the data to the research propositions had two facets. One was the characteristics of the Li Lake area. These characteristics included those remembered and recorded in general, those emphasised with by clients, those identified and interpreted by international practitioners, and those recognised in various local and international awards. The other facet was the approaches for gaining local understanding, including the ways international practitioners studied the area and methods they engaged diverse local stakeholders. Because there were two international design

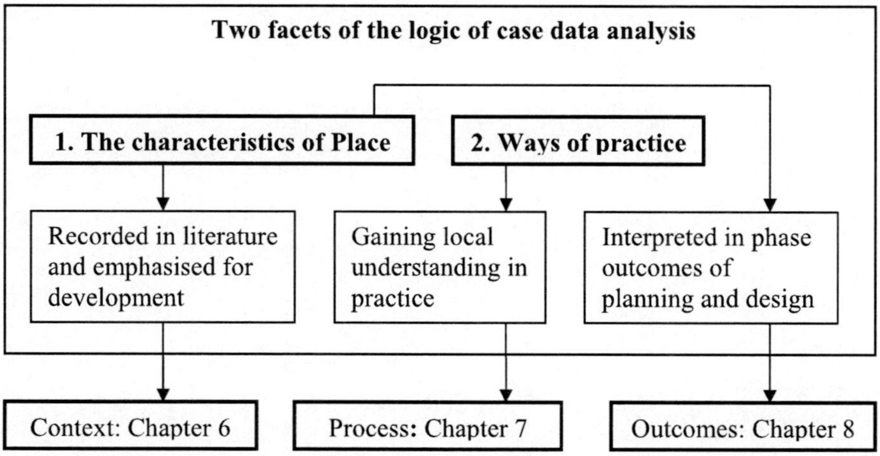

Fig. 3.2 Analytical approaches to the case data

teams involved independently in this case project, the analysis of the process and outcome was aimed at each team. Figure 3.2 explains the analytical approach of the case study. The analyses of the three units are presented in Chaps. 6, 7 and 8.

References

Cantor, S. L. (1997). *Contemporary trends in landscape architecture*. New York: Wiley.

Fink, A., & Kosecoff, J. (1998). *How to conduct surveys: A step-by-step guide*. Thousand Oaks, California: Sage Publications.

Fowler, F. J. (2002). *Survey research methods*. Thousand Oaks, California: Sage Publications.

Francis, M. (2001). A case study method for landscape architecture. *Landscape Journal, 20*(1), 15–29.

Rea, L. M., & Parker, R. A. (2005). *Designing and conducting survey research: A comprehensive guide*. San Francisco: Jossey-Bass.

Strauss, A. L., & Corbin, J. M. (1998). *Basics of qualitative research: Techniques and procedures for developing grounded theory*. Thousand Oaks: Sage Publications.

Yin, R. K. (2003). Case study research: Design and methods. Thousand Oaks, California: Sage Publications.

Chapter 4
Australia's Supply of International Landscape Architectural Services

Abstract This chapter investigates Australia's supply of international landscape architectural services, and especially to China. This chapter is the counterpart of Chap. 5, which examines China as a site of demand for international practice in landscape architecture. Together, the two chapters explain the larger context for the case study of the work provided by international design teams for the urban development of the Li Lake area in Wuxi, China, which involved Australian landscape architects as key players. This chapter comprises three major sections. Section 4.1 examines Australia's geographical and cultural position in the Pacific periphery and specifically its relationship with Asia and China. Section 4.2 surveys the international landscape architectural services provided by Australia during the period from 1995 to 2007, with a focus on China. Section 4.3 discusses challenges to and strategies for Australian practitioners providing landscape architectural services to China.

Keywords Australia's landscape architectural services in China · Business characteristics · Work characteristics · Challenges · Strategies

4.1 Exports to Asia and China: Australia's Geographical and Cultural Position

Australia's supply of international landscape architectural services is underpinned by its geographical proximity to its Asian neighbours. Wingrove (1999, p. 245) claimed that in the 1990s, links between the countries on the Pacific periphery were promoted by rapid economic growth and new trading patterns. This periphery is bounded by Asia and Australia to the west and by the Americas to the east, suggesting a close relationship between the coast lines of Asia and Australia. The literature on Asia-Australia relations (Andrews 1992, p. 271; Australia Parliament Senate Foreign Affairs Defence and Trade References Committee and Forshaw 1996; Kendall 2005, p. 166) also suggests that the economic nexus of Asia and Australia results not only from Australia's geographical proximity to its Asian neighbours, which is much closer than that of America, but also from Australia's ambiguous position among the world powers 'between Europe (or the United States) and Asia (or the Pacific)'

Y. Zhang, *Place Making in International Practice of Landscape Architecture*, Advanced Topics in Science and Technology in China 64,
https://doi.org/10.1007/978-981-16-2442-1_4

and its political approach to security and balance. The above studies suggest that Asia has been a major destination for Australia's supply of services internationally, including landscape architectural services.

The emerging market in China for Australia's service exports surged in late 1993 and was further promoted by the financial crisis in Southeast Asia in 1997. In 1996, the Australian Parliamentary Senate Foreign Affairs, Defence and Trade References Committee (APSFADTRC) and Forshaw (1996, p. 55) identified the booming economic relations between Australia and China in late 1993. They pointed out that Sino-Australia economic relations were the core of the overall relationship between the two countries. More specifically, their research indicated a shift in the composition of Australia's exports to China, which were marked by 'a decrease in the proportion of primary products and an increase in the proportion of manufactures and services' (1996, p. 56). Their research cited the prediction of Austrade,[1] which estimated that by the turn of the century, the expansion and diversification of the two-way trade would make China the second largest export market (following Japan) for Australia and the fastest growing regional market (1996, p. 57).

Kendall (2005) argued that the 1997 crisis promoted Australia's relations with East Asia, which is evident in the invitation for China, Japan and South Korea to join the Association of Southeast Asian Nations (ASEAN), and the formation of 'ASEAN-plus-three' (Kendall 2005, pp. 163–164). It could be argued that the shift of the exportation of services from Southeast to East Asia, implies that trade would be occurring between nations with greater cultural differences, as East Asia has a shorter history of western colonisation. In 2000, in his speeches in Beijing, Downer, who at the time was serving as the Foreign Minister of Australia, said that Australia could engage Asian countries through 'practical regionalism' (2005, p.163). Downer meant that while Australia and Asia have cultural and ideological differences and do not share a common history and cultural identity, they are 'bounded together by geography', and both can find practical ways of working together to achieve their mutual objectives (Kendall 2005, pp. 163–164). This further shows Australia's willingness to export services to its Asian neighbours and to expand its market in China.

Australia's expanding markets in Asia and China, are manifest in Australia's export of architectural services. The *Review of Legislation Regulating the Architectural Profession* (Productivity Commission 2000, p. 59) reported that the Department of Foreign Affairs and Trade (DFAT) identified Asia as being an important destination for the export of Australian architectural services, saying that 'the Australian architectural profession has identified Indonesia, Malaysia, Singapore, Hong Kong and Thailand as their key markets in Asia'. It also cited the DFAT's report as saying that 'China, Vietnam and the Philippines have been identified as emerging growth markets' (2000, p. 59) and that the export of architectural services to Asia underlies 'an important part of some architects' revenue, with the potential for future

[1]Austrade—the Australian Trade Commission, which is the Australian Government's trade and investment development agency that helps Australian businesses to succeed in international trade and investment.

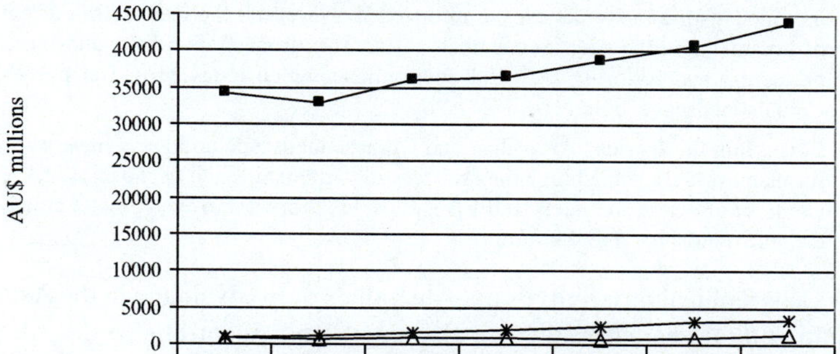

	2000	2001	2002	2003	2004	2005	2006
—■— Exports of services to the world ($ millions)	34284	32969	36023	36382	38672	40656	43873
—△— Exports of architectural services to the world	986	780	816	804	723	991	1268
—✳— Exports of services to China	792	1209	1558	1923	2521	3029	3332

Fig. 4.1 Australia's international trade in services. *Source* Australian Bureau of Statistics (2007)

expansion, particularly for medium to larger-sized practices' (2000, p. 59). Although this review was specific to architectural services, the Royal Australian Institute of Architects (RAIA) suggested that these services may also include those provided by landscape architects (2000, p. 41).

The dramatic increase in Australia's provision of architectural services to China is also explained by figures presented by the Australian Bureau of Statistics (ABS) in 2007. To illustrate, Fig. 4.1 shows that the income earned from Australia's overall exports of services increased during the seven-year period from 2000 to 2006 but decreased in 2001. The export of architectural services also increased 1.28 times, from AU$986 million in 2000 to AU$1268 million in 2006, and decreased in 2001 and 2004. Regarding China specifically, the available figures provided by the ABS (2007) show Australia's overall export of services to China increasing from AU$792 million in 2002 to AU$3332 million in 2006, or nearly five times. The exports of overall services to China increased each year. The increase was most dramatic in the year 2001, when the figure was 1.53 times that of the previous year.

The continuous increase of service exports to China suggests that the increase of Australian landscape architectural services and influences there. This is evident in the comments of the national president of the Australian Institute of Landscape Architects, Mark Fuller, in 2006, when China became a member of International Federation of Landscape Architects:[2]

> China is one of the most important countries in our region and one in which many Australian Landscape architects are working. The influence of its human and natural history has always

[2]http://www.aila.org.au/media/pages/ifla-china.htm, retrieved on 30 June 2010.

> provided inspiration for landscape architects, however it is the scale of the current challenges of urbanisation, the need to establish sustainable urban frameworks and the challenge of protecting and restoring the rural environment which today provides the challenges for China's landscape architects.
>
> In tackling these issues Australian and Chinese landscape architects have worked together for many years, establishing many successful partnerships and friendships. AILA welcomes China's landscape architects to IFLA and we look forward to even greater collaboration and deeper friendships into the future.

Geographical proximity is not Australia's only advantage in the Asian market; its multicultural characteristics are also important. According to the ABS (2008), the national census in 2006 found that Chinese was the most extensively used second language used in Australian homes, at 2.3% (1.1% Mandarin; 1.2% Cantonese). The second was Italian, which made up 1.6%, and the third was Greek, which made up 1.3%. The ABS (2008) also found that since 1996, people of Chinese origin have contributed the second-most number of people to the increase of population in the countries where Australian residents were born. This number stood at 96,000 in the year 2006, while New Zealand was ranked first at 98,000 and India was ranked third at 70,000. In addition to the Sino-Australia links facilitated by Cantonese in Australia, in 1989, Chinese immigrants in Australia started to come from various areas of mainland China including Shanghai and Beijing (Fung and Chen 1996). It could be argued that these Australian-Chinese contribute further to the expansion of Australian planning and design services in China from southern coastal regions of Hong Kong and Canton Province to the inland.

The geographical and cultural position of Australia, discussed above, underpins two common characteristics of its supply of international landscape architecture practices: first, Asia is the major destination of Australia's international design, which creates challenges stemming from the cultural and ideological differences. Second, Australia's international practices are aided by immigrants from Asian countries (2005: 164), who may transmit intangible influences, such as cultural flexibility and an outward orientation towards Asia. These two characteristics are also implied in the following survey findings.

4.2 Australia's Landscape Architectural Practice Internationally and in China

In August 2007, a survey was conducted of 272 landscape architectural firms that registered their businesses on the website of the Australian Institute of Landscape Architects (AILA), a member of the International Federation of Landscape Architects (IFLA) and the registering body for the profession in Australia. The survey focused on Australia's supply of international landscape architecture practices during the period from 1995 to 2007 and its supply to China. The survey was carried out through three themes of inquiry. The first was concerned with the characteristics of the firm (size, location and income generated) providing landscape architectural services internationally. The second inquired into the landscape architectural services that these firms had provided for China, including the characteristics of the work itself (types and location of projects undertaken, source of work, types of services provided, key contacts and relationships). An article (Zhang and Bull 2008) was published in October 2008 reporting on the findings from the two themes of inquiry and the data that they had generated. The third theme of inquiry was concerned with the challenges of and strategies for providing landscape architectural services for China. The results of this inquiry are discussed in Sect. 4.3. The full list of questions is included in the appendix.

Of the 272 survey questionnaires mailed, 104 (approximately 40%) responses provided the basis for the following data and its analysis. Of the 104 respondents to the survey who practiced, 51 (approximately 50%) confirmed that they had practiced internationally during the period from 1995 to 2007 (Fig. 4.2). This proportion exceeds the ABS's finding, which reported that 14% of practices were exporting services internationally in the years between 1997 and 98 (Productivity Commission 2000, p. 59). Further, 37.5% of the respondents were involved in projects in China, which highlights the significance of China as a client. Australia's exports of landscape architectural services also increased in this period, with 33 firms having practiced internationally prior to 1995 and 47 in the period between 2005 and 2007.

International practice including China is not confined to landscape architectural practices in a single Australian city or location, but appears to be spread across states, concentrating on the three states along the Pacific seaboard, Queensland (QLD) to the north, New South Wales (NSW) and Victoria (VIC) to the south. Landscape architectural practices in NSW appeared most active in providing services for China (Fig. 4.3).

The Australian Bureau of Statistics (ABS) identified small practices (under 20 employees) as dominating the architectural services businesses, and medium (20–49 employees) to big (50 plus employees) as being under-represented. To illustrate, at 30

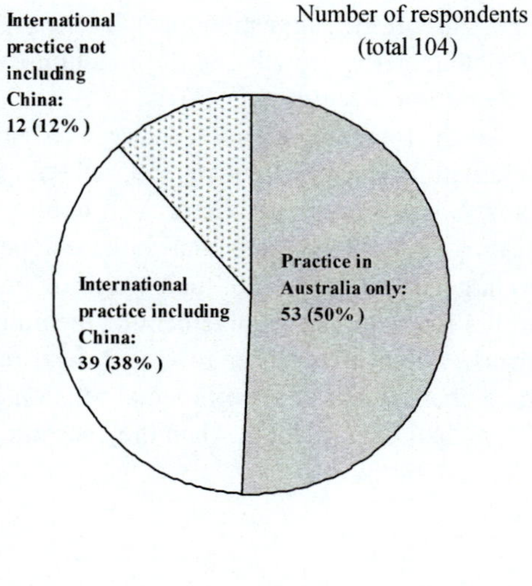

Fig. 4.2 Business characteristics—involvement in international practice. *Source* The questionnaire survey

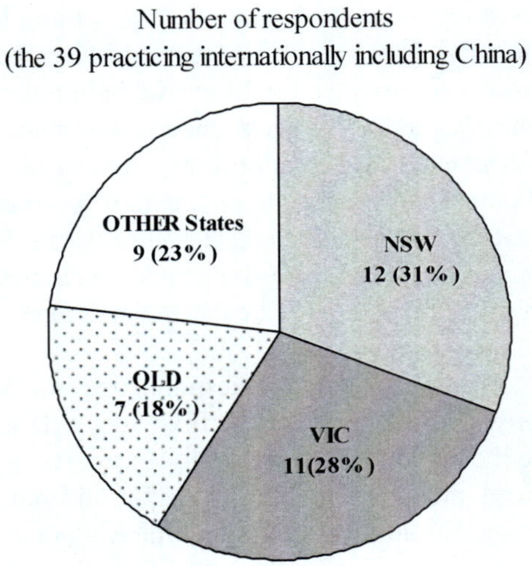

Fig. 4.3 Business characteristics—location in Australia. *Source* The questionnaire survey

June 1993, the number of medium-to-big sized architectural firms was 2% of the total number of the business in architectural services (McLennan 1995, p. 1). Such a characteristic of size is consistent with that of landscape architectural practices providing services internationally over the decade between 1995 and 2005. In comparison with the number of architectural businesses, which was 4409 surveyed at 30 June 1993 (McLennan 1995, p. 1) the number of landscape architectural businesses under survey was 272 in August 2007. However, the proportion of medium-to-large sized practices within landscape architecture was higher than that of architecture (18–27% compared with 2%).

In answer to question one (number of employees), respondents reported that the involvement (numerically) in international practice was dominated by small firms (72% in 1995 and 58% in 2005). The number of medium and large firms increased during this period: medium practices increased by 50% (from a relatively small base) and the number of large practices doubled at the expense of small firms. These data support the suggestion made by the Productivity Commission (2000) that there is potential for growth for the medium-to large-sized firms in this sector in the exportation of services.

Of the three major states, the reduced involvement of smaller firms in providing international services during 1995–2005 appears to have been a consistent trend in New South Wales and Victoria, in contrast to the growth in Queensland in 2005 (Fig. 4.4). In New South Wales, there appears to have been increased involvement in large, at the expense of medium firms. In Victoria, there was a significant growth in the international involvement of medium-sized firms. The involvement of small firms, which grew faster in Queensland than in other states, can be attributed to the enhancement of governmental relationships between Queensland and Shanghai Municipal City in 2004, when they became sister states (Beattle 2005).

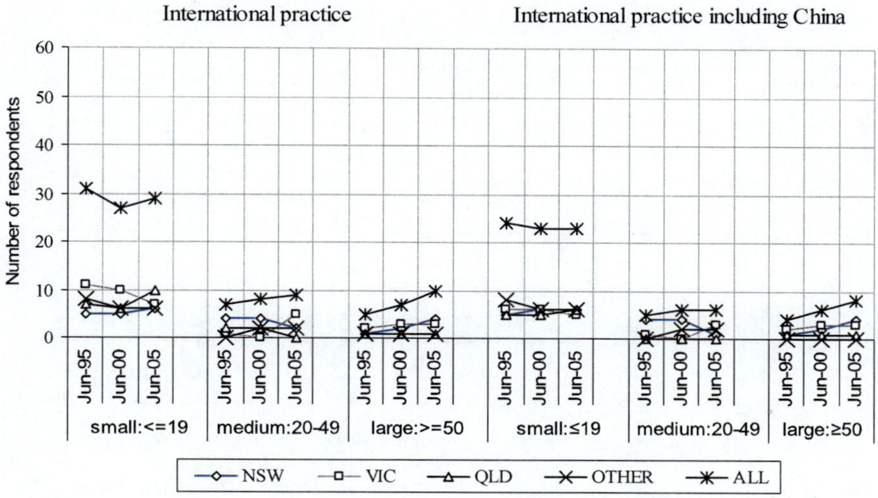

Fig. 4.4 Business characteristics—size of firms. *Source* The questionnaire survey

Of international practices that provided services to China, respondents from small-sized firms comprised less than half (around 43%) in the decade between 1995 and 2005. Of medium-sized firms during the same period, more than two-thirds practiced in China (between 67 and 75%). Nearly all of the large-sized firms practiced in China (between 80 and 86%), indicating that China was a principal destination for medium-to-large Australian firms exporting landscape architectural services during the time. This also suggests the relatively large scales of the projects being undertaken in China, which could apparently only be handled by large landscape architectural firms. The research of the APSFADTRC and Forshaw (1996, pp. 83, 107) found that China tends to favour Australia's large providers rather than its small-to-medium-sized businesses (less than 20 employees for export services and less than 100 for export manufacturing), who reportedly 'find it difficult to both generate venture capital for projects in China and win approval of small orders from Chinese authorities'. In contrast to the APSFADTRC's findings, the survey revealed that medium-sized firms were involved in Australia's exportation of landscape architectural services and appeared to be active in the Chinese design market.

In terms of income (question two), the analysis is based on the principal respondents' estimates rather than on actual statistics. These estimates, however, do reveal specific trends across time. First, the majority of the 51 respondents practicing internationally worked in small-sized firms (72% in 1995–2000 and 60% in 2000–2005), and for most of these small-sized firms, income from international work comprised less than 10% of their annual total (Fig. 4.5). This suggests that international work was not a major income earner for small-sized firms. Second, for medium-sized firms, the figures showed that in the period between 2000 and 2005, there was an increase of firms for which their income from international work rose to 30%. Finally, for large-sized practices, those whose international income was less than 10% no longer

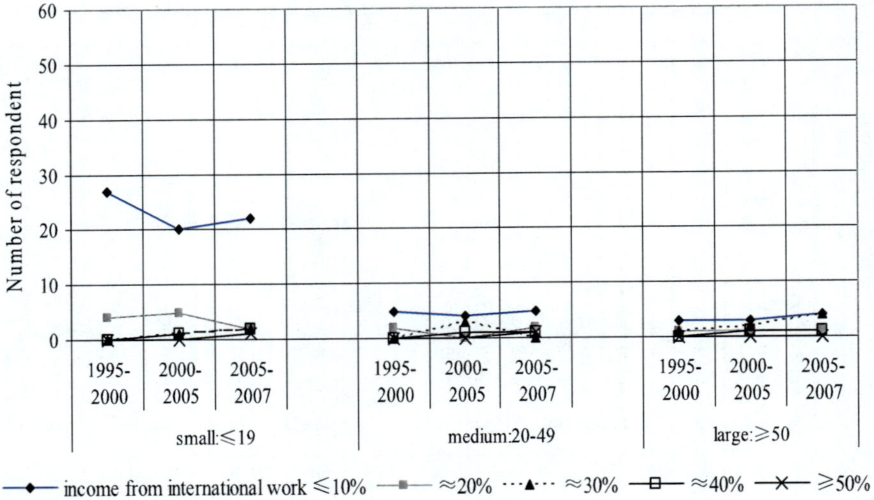

Fig. 4.5 Business characteristics—income from international work. *Source* The questionnaire survey

dominated in 2005–2007. By contrast, in this recent period, there were a similar number of big-sized firms for which their international income rose to 30%.

The answers of the 39 responding practices providing landscape architectural services to China to the third question (contribution of projects in China to the total income from international projects) shows that small-sized firms comprise more than two-thirds (between 73 and 88% during the period from 1995 to 2007) of those whose income from work in China was less than 10% of their income from international work, with a significant decline in the period from 2000 to 2005 (Fig. 4.6). Of the

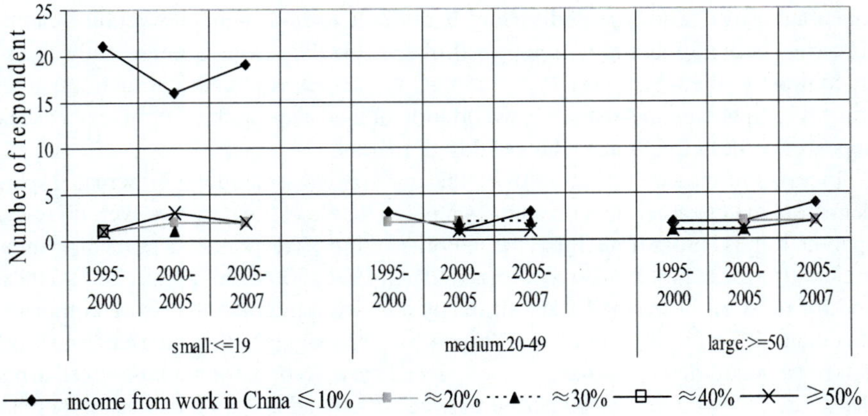

Fig. 4.6 Business characteristics—income from work in China. *Source* The questionnaire survey

responding firms that provided services for China, four large-sized firms experienced growth in almost all of the income categories, yet that growth was from a relatively small base. A comparison of Figs. 4.5 and 4.6 shows that income from work in China changed more dramatically than income earned from the total international work over the same period.

The above figures indicate that although nearly half of the practices provided landscape architectural services internationally, the majority earned less than 10% of their overall income from that work during this period. While the Asian economic crisis in 1997 caused a decline in fees earned overseas in architectural services (Productivity Commission 2000, pp. 58–59), the decline appears only to have affected small- and medium-sized landscape architectural firms who earned less than 10% of their income from international work. In contrast, there was an increase in medium- and large-sized firms who earned a greater proportion of international income in their annual total during this period. This phenomenon may be related to the economic growth that occurred in China during this time and the consequent opportunities for these firms to provide services for large-scale projects there.

Participants were also surveyed to gain insights into the kind of services they provided for China (whether urban or rural), and their basic approaches, the phases in which they were involved (pre-design, schematic design, design development, construction, etc.) and the type of interactions they had with the local working community (who and how).

In response to the fourth question, which investigated the types of projects the firms were involved in, respondents chose and ranked the project types that they undertook in China between 1995 and 2005. The data show that their services concentrated in the urbanising domain and could be categorised into five major areas: urban public space, residential districts, urban precincts, inner urban, and the urban fringe (green fields in the outskirts of cities). Between the first five years and the second five years of the decade (Fig. 4.7), projects in residential districts doubled, and projects in public spaces and the urban fringe tripled. While only 3 of the 39 respondents identified themselves as having done work in the urban fringe in the first five years of the decade (1995–2000), this figure increased to 11 in the second five years (2000–2005). This suggests that in the urbanising domain of Australian landscape architectural practice in China, the growth of Australian landscape architectural services in the urban fringe has been the quickest, revealing that the urban fringe in China was an area where landscape architectural services were increasingly in demand.

The number of respondents indicating they had completed projects in scenic rural areas also doubled over the decade (Fig. 4.7). Other types of projects were also undertaken by respondents, including waterfronts, universities, highways, resorts, hospitals, recreational facilities, zoos, commercial areas, tourism facilities, educational facilities and the Australian embassy. This suggests a rich variety and a significant increase in project types over the decade.

Comparing the type of work undertaken by firms from different Australian states in residential districts, urban precincts, and urban public spaces in China where Australian landscape architectural practices have accumulated, it shows changes that occurred between 1995 and 2005. In the second five years, New South Wales

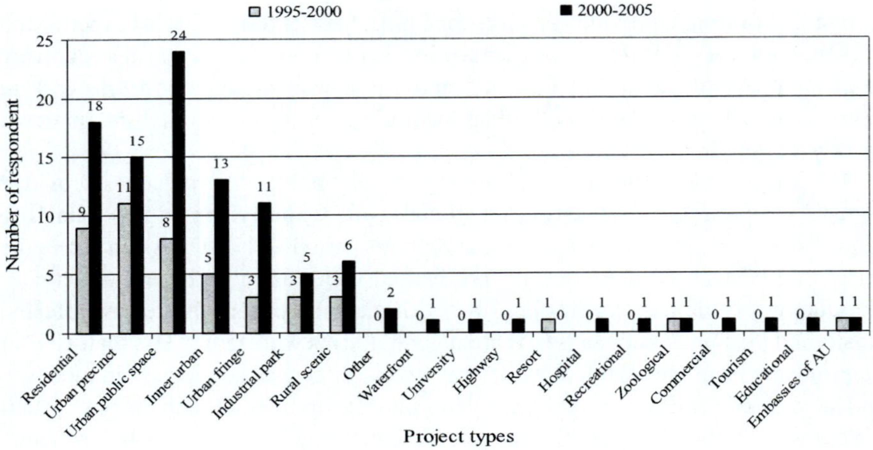

Fig. 4.7 Work characteristics—project types. *Source* The questionnaire survey

took first place over Victoria and the other states, focusing on urban public space (Fig. 4.8). Queensland also appeared much more active during this same period. Such an increase in projects associated with China's public domain also suggests the enhancement of relationships between Australian practitioners and Chinese local governments. Of the three major states, New South Wales and Queensland appeared to be more active in designing urban public space in China; practitioners from Victoria worked primarily in residential areas.

Participants were also asked to nominate the ways (question six) that they interacted with various local supporters in typical services they provided that characterise landscape architecture over the decade between 1995 and 2005. According to the 39 responding firms providing services for China, overall the more highly-skilled

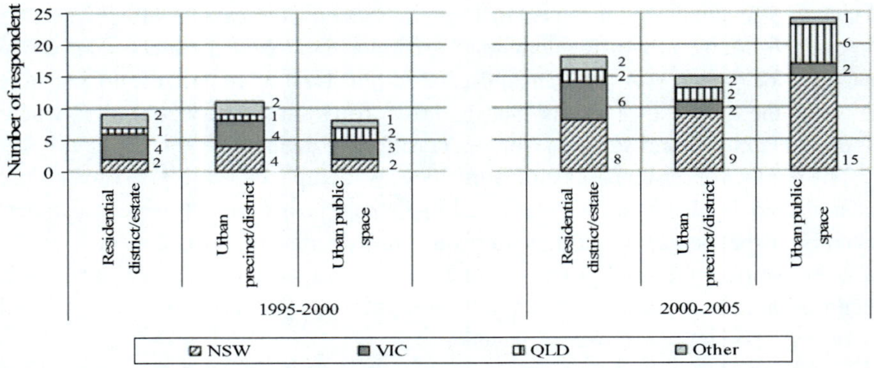

Fig. 4.8 Work characteristics—project types/locations of Australian firms. *Source* The questionnaire survey

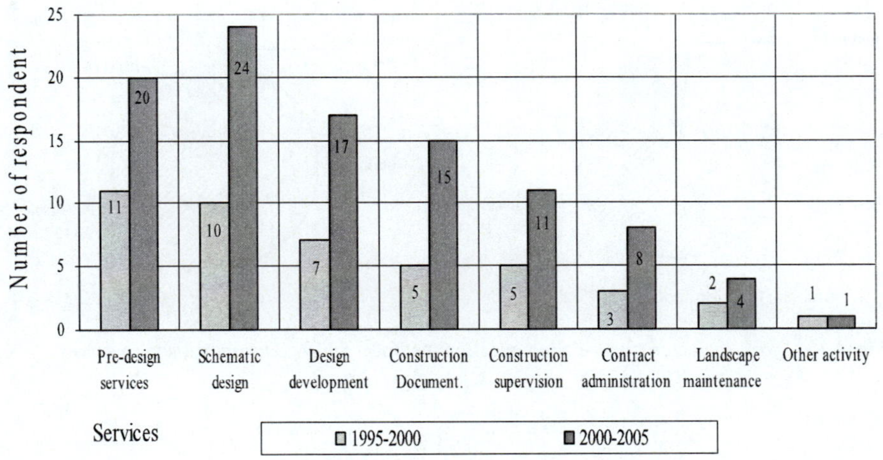

Fig. 4.9 Work characteristics—service category. *Source* The questionnaire survey

phases of pre-design (brief preparation, reports, feasibility studies, master planning and strategies, etc.) and schematic design dominate. From design development to construction documentation, the involvement of Australian landscape architectural firms was progressively reduced, presumably because these were increasingly provided by local practitioners (Fig. 4.9). All categories of services, however, were subject to growth across the decade. The service of providing construction documentation increased the most over the second five years of the decade, growing at three times the rate of the previous five years. The rate of growth in this area was followed by an increase in providing contract administration and design development. The provision of pre-design services increased the least, falling to second place in the last half of the decade.

The ways in which Australian practitioners engaged with various local supporters of their work in China, such as local clients, professionals, and builders, varied across the respondents. In general, periodic visits and e-communication were identified as the main methods of engagement and were used more often than the method of establishing branch offices and posting staff there. However, establishing local branch offices did increase. When 11.8% of firms had branch offices in the first five years of the decade, the figure rose to 16.2% in the second five years (Fig. 4.10).

There were noticeable changes between the first and second five years of the decade in the ways Australian practitioners and landscape architects engaged local stakeholders during the different phases of providing landscape architectural services. In the first five years, periodic visits were used equally during the pre-design and schematic design phases (Fig. 4.11). In the second five years, this method was used more during the pre-design services phase than during the schematic design phase. In the second five years, along with their increased general use, branch offices were also increasingly used as bases to coordinate contract administration and landscape

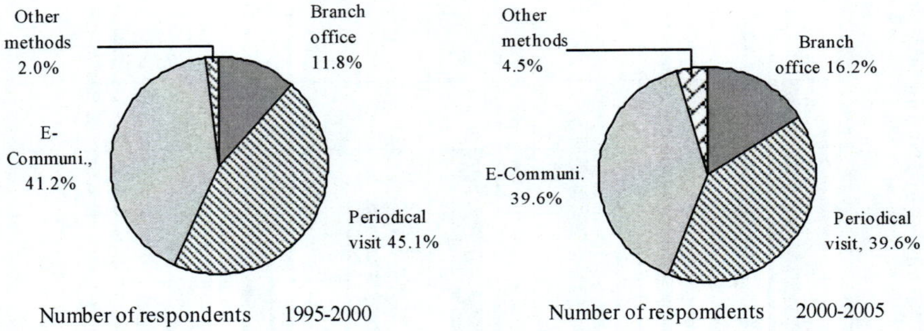

Fig. 4.10 Work characteristics—communication methods. *Source* The questionnaire survey

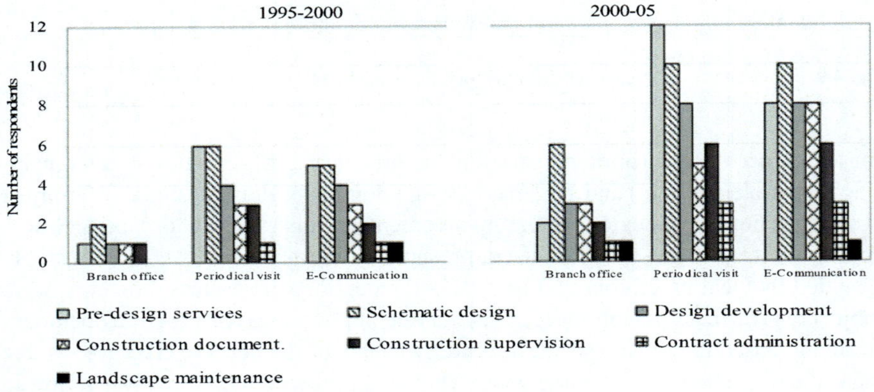

Fig. 4.11 Work characteristics—communication/service category. *Source* The questionnaire survey

maintenance. Branch offices became progressively more important in the later stages of the landscape architectural services provided in China during this period.

In response to question five, which inquired how and where work was sourced, the 39 respondents indicated three key sources of their work. The first was prior relationships (33% of the 39 respondents in 1995–2000, 46% in 2000–2005). This means that the work they did in China came from previous cooperation with other consultants, such as architects (Fig. 4.12). The second was personal recommendations (26% of the 39 respondents in 1995–2000, 43% in 2000–2005), such as recommendation from previous or current clients. The third was joint projects with other consultants (18% of the 39 respondents in 1995–2000, 38% in 2000–2005), again, such as architects. Openly sourced information (publicly available sources including media like newspapers, magazines, radio, television and company websites) that reported the firms and their work was much less influential.

Two sources, 'prior relationships' and 'personal recommendation' were ranked by respondents as being of similar importance in their ability to receive work from

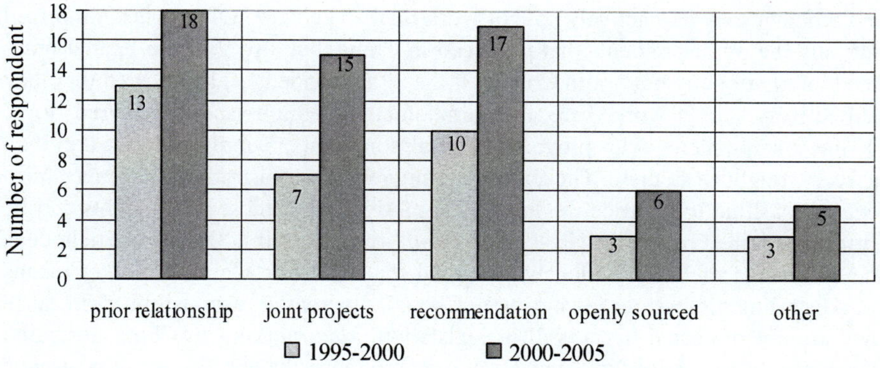

Fig. 4.12 Work characteristics—sources of projects. *Source* The questionnaire survey

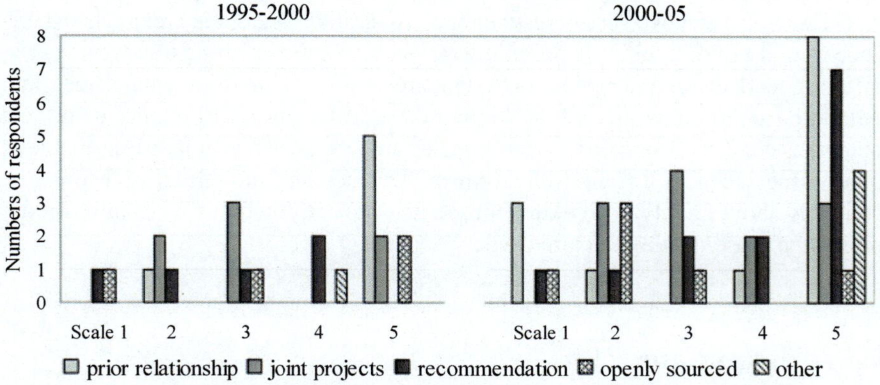

Fig. 4.13 Work characteristics—importance of sources of projects. *Source* The questionnaire survey

China in the second five years of the decade (2000–2005). While 'prior relationships' was used more in the first five years, 'personal recommendation' became increasingly important (Fig. 4.13). This suggests the significance, for Australian landscape architects, of building trust with individual Chinese clients and local stakeholders, and developing inter-personal skills.

4.3 Challenges to and Strategies for Practicing in China

The third theme of the survey (questions seven and eight) concerned Australian practitioners' opinions about the services they provided for China, including the challenges they confronted, the strategies they employed to address these challenges, and whether they had actively sought work there. The responding firms comprised not only the 39 respondents that practiced in China, but also 9 of the 12 that practiced internationally but not in China. Of the 39 respondents, 11 indicated that they had actively sought work there, including small-, medium- and large-sized firms. Of the 9 respondents who provided general comments, 3 indicated that they had actively sought work there. The above data suggests that Australian landscape architects are willing to practice in China. The motivation of these respondents can be further explained by the article: *Looking both ways* (Nevill 2006). The article cites the comments of Mark Fuller, who argued that what the Chinese market means for Australian professionals working in China is 'increased responsibility', because they are not protected by 'a wall of legislation', and working in China 'tests and develops professional skills'. Such a claim also suggests that the process of landscape architectural practice in China is much more dynamic than that of Australia.

At the same time, the more dynamic process of practice in China and increased responsibilities of Australian landscape architects mean greater challenges. According to the above survey respondents, Australian landscape architects experienced a range of difficulties in providing services in China: creating trust, extra costs, differing work ethics and professional standards, different environmental values, and unfamiliarity, to name just a few. Among these, extra costs and quality of project outcomes were the two major concerns and ultimately these two dimensions included other issues such as different environmental values and unfamiliarity. Moreover, concerns about quality design and consequent project outcomes were raised more often than those relating to extra costs.

4.3.1 Extra Costs

The extra money, time and energy spent working on projects in China over that required for equivalent local projects were the major sources of the extra costs. One respondent from a firm in Sydney indicated the failure of Chinese clients to pay fees promptly and the double taxation of the services by both China and Australia. From such a perspective, providing landscape architectural services, whether local or international, is a form of business confined by income and costs. The survey reveals that four strategies have been used to reduce extra costs: first, establishing local offices; second, building local associations; third, being a large-sized firm; and lastly, working on large projects.

To illustrate, some respondents suggested that the establishment of local offices was a prerequisite for gaining knowledge of the Chinese legal system and business culture, saying that without this knowledge, working in China may not be sustainable. Other respondents suggested that building local associations, perhaps with local practitioners, could improve their relationship with the local government and with private individuals, which can help Australian practitioners to understand the often opaque legal system in China. In addition, having reliable Chinese partners was seen by some respondents as helpful in building business relationships and trust with Chinese clients and in understanding local business operations and standards. A respondent, who was among the 9 practicing internationally but not in China, commented that only large firms would risk working there. Some respondents also suggested that local branch offices are more affordable to large firms and that large projects had a greater influence on the local area and nearby regions and therefore brought in more projects.

The above discussions suggest that the four strategies for reducing extra costs are correlated. This, however, highlights the greater challenges of small-sized practices. Of those small-sized responding firms having provided services in China, one indicated that they had to rely more on larger practices and worked as subcontractors. They therefore had less influence on project outcomes. One respondent indicated the need for international recognition, which is often difficult to achieve for small-sized firms. Such comments explain why a majority of small Australian landscape architectural firms chose to provide services only within Australia.

In spite of the greater challenges caused by small size, of the 11 responding firms that provided services in China and claimed that they had actively sought work there between 1995 and 2005, six were small-sized firms. Of the six firms, two grew from small sized in 2000 to medium sized in 2005, and the remaining three were large firms from 1995 to 2000. In addition, three responding practices that practiced internationally but not in China and claimed that they had actively sought work in China were all small firms and remained so across the decade. One respondent, whose firm had grown from medium to large by 2005, indicated that involvement in projects in China 'allows opportunities to develop' and 'consolidate firm business relationships'. All of these suggest that quite a few small Australian landscape architectural firms regarded the involvement in projects in China as opportunities to develop their businesses. This also suggests that in the initial participation of small or medium Australian landscape architectural firms in the Chinese design market, the most useful strategy for reducing extra costs is to build local associations. The other three strategies (establishing local offices, being a large firm and undertaking large projects) can only become possible after these smaller firms have developed.

These small-sized firms were also put at a disadvantage by Australian policy. APSFADTRC and Forshaw (1996, pp. 105–107) found that small- to medium-sized Australian enterprises have had key roles to play in Australia's economic engagement with China. However, they have had more barriers and face more difficulties than larger firms, both from China and Australia. APSFADTRC and Forshaw (1996, p. 105) suggested the difficulty from within Australia was that 'government assistance to the private sectors operating in China biases towards larger companies'. Their research (1996, pp. 80–81) also indicated the risks taken by Australian service providers when looking for a foothold in the Chinese market, which were usually caused by insufficient knowledge of the business culture in China. That research pointed out that business issues in China, such as the need to spend time building personal contacts, negotiating and concluding deals, are very different from those in Australia and many other export markets.

The abovementioned differences in business cultures were further commented on by respondents to the survey. They revealed more issues faced by Australians relating to business culture in China, including clients' reluctance to discuss problems, the lack of transparency in decision-making, a preference for bartering and sensitivity to prestige. These issues were seen by Australian practitioners as having an influence on project outcomes. These issues also suggest that the extra energy and time costs of working in China were based on the cultural differences between Australia and China, especially business cultures. Such cultural differences seem to be greater than those between Australia and other Asian countries, such as those in Southeast Asia. This greater difference is evident in the comments of one respondent, who indicated his preference for working in Southeast Asia and claimed that this area was 'more English-based than China' and 'imply more common links'. Therefore, cultural differences were fundamental to the extra costs of time and energy noted by Australian landscape architects who practiced in China. Compared with these issues, the extra monetary costs, such as the double tax, are less fundamental. This is evident in the fact that 37.5% of Australian landscape architectural practices have provided services for China, despite international income not being a major income earner for their practices. Further, some practices, although not yet involved in working in China, have been actively seeking work there.

4.3.2 Quality Project Outcomes

Another major concern Australian practitioners have about providing landscape architectural services in China is the quality of project outcomes. The major challenges raised by the respondents comprised three aspects: clients, landscape architecture as an industry and profession in contemporary China, and the cultural differences between Australia and China, which was seen as underpinning them all.

To illustrate, challenges related to interacting with clients included the lack of clear design briefs, unrealistic expectations, conflicting expectations between Western styles and Chinese traditions, and clients' lack of support for practitioners' proposals for sustainable development. Challenges for landscape architecture as an industry in China included limitations in the provision of landscape materials, such as plants; low local construction skills and standards; uncertain professional standards; lack of a sense of the value of quality design; and 'the lack of environmental and cultural critics in Chinese project/practise'. Challenges related to cultural differences were referred to as perceived 'culturally different local values', the perceived 'limited understanding of cultural nuance', and 'language barriers'.

In response to these challenges, the respondents proposed various strategies. To meet the challenges from Chinese clients, the strategies of Australian practitioenrs included first, taking on projects implying more opportunities for developing partnership with local clients. The second strategy was creating better mutual understanding between clients and Australian practitioners, including educating themselves about local values and conventions and facilitating clients' understanding of the values of quality design. One respondent commented that promoting such mutual understanding can change clients' unrealistic and conflicting expectations, and can, in the end, help Australian practitioners obtain support from them for sustainable design ideas.

Strategies developed to handle the challenges associated with landscape architecture as an industry and profession in China included maintaining influence over the construction process to secure the quality of the built landscape. According to respondents, this includes winning the right to select local subcontractors and providing training to improve local construction skills. One respondent of a firm in Hobart commented that China had 'limited values and respect for their own design skills and heritage values of place'. This respondent suggested that education programmes should be provided to improve local environmental awareness, and to have cooperation with local counterparts.

The strategies proposed by respondents to meet the challenges associated with cultural difference, however, were ambiguous, except for the establishment of local associations. One respondent from a firm in Melbourne, referring to the conflicts between clients' expectations for designs that combine both modern Western and traditional Chinese elements, said only the Chinese can find the balance. Some respondents suggested that Chinese clients' preferences for modern Western styles followed from the worship of unsustainable Western paradigms.

The strategies proposed by respondents also presented contrasting approaches. To illustrate, although large-sized firms and large projects were commonly regarded as the means of achieving financial stability necessary for establishing an international practice, one respondent from a large firm in Sydney proposed that reducing the volume of work was a way to ensure better outcomes. Strategies of Australian practitioners also included assigning staff with Asian backgrounds to the project to bridge the conflicting business cultures and develop relationships with clients that go beyond business. In contrast, there were also suggestions about how to cater to clients' preference for Western styles, such as deploying staff with Western cultural backgrounds to carry out the conceptual work.

The suggestions of respondents to provide training and to encourage critique of the practice and the projects as well as respect for China's cultural values show the good intentions of Australian practitioners and their desire to provide quality services. These, in turn, reflect the problems with project outcomes that have concerned them. Such concerns were evident in the comments of a respondent from a firm in Melbourne, who indicated:

> Communication is obviously the biggest challenge dealing with Chinese clients[,] however we have noticed their English improving. Interpreters are not necessary as most professionals have basic English[,] however many constructed detailing is misunderstood[.] I have concerns about the obvious the use of foreign design skills which are so appealing to the Chinese wanting to base their project design on the western tradition (I have obvious philosophical issues with this approach)[,] but once these [international] consultants provide a master plan or concept plan[,] they then proceed with design development, contract documentation and contract administration locally.

This respondent provided an unsuccessful example of a project in China, saying (Fig. 4.14):

…while the concept of tactile paving is considered appropriate at a master planning level but may not be understood and documented correctly which is usually undertaken by the local design firm (cheaper fee!!) but has not been designed as intended thus in this example runs directly into trees, bicycle racks and other street furniture…China is such a huge market and after our initial experience we maintain a philosophy that anything is possible in China.

Fig. 4.14 Practice outcomes—an unsuccessful example. *Source* The questionnaire survey

Overall, strategies for improving mutual understanding with clients and maintaining control during the construction phase of the project were expressed more explicitly than those about bridging the gap between cultural differences. Establishing local offices was frequently cited as a sound approach, suggesting that such an approach was generally recognised as useful. A respondent from a medium-sized firm in NSW indicated that a long-term and visible regular presence in the local area was essential for achieving quality project results. This emphasis on establishing local offices suggests that quality results are associated with the localisation of the practice.

The APSFADTRC and Forshaw (1996, p. 81) has suggested that service providers capitalise on the expertise and experience of staff returning to Australia from postings in China. These individuals have critical first-hand knowledge of the Chinese market and knowledge acquired through long-term familiarity with the area. Australia's exporters of landscape architectural services to China have responded by establishing local offices. However, the APSFADTRC and Forshaw also found a lack of networking and co-operation among the Australian business community exporting to China. This suggests that, for Australian landscape architects, a further step is required to reduce the costs of providing services to China and to improve the quality of project outcomes: constructing a network among the local offices of the Australian landscape architectural practices in China.

4.4 Conclusion

The survey showed that Australia's identification with Western cultures posed a fundamental challenge to Australian landscape architects providing services to Asian countries. Such challenges have been greater in providing service in China than to Southeast Asia, as cultural and ideological differences between Australia and China are greater than those between Australia and other Asian countries. Australian landscape architects generally recognised such socio-cultural differences, but with a focus on those in business cultures, rather than in communities and ways of life.

The survey also found that small Australian landscape architecture businesses had more initiative than large ones to seek work in China. The decrease of small-sized firms and the increase in medium-to-large-sized ones between 1995 and 2005 suggests a shift in practice structure in that period, some of which can be associated with the growth that many firms achieved by working in China. The high proportion of Australia's exports of landscape architectural services to China suggests the importance of China's urban development to the profession of landscape architecture in Australia.

The survey findings suggest that income was not the principal reason for Australian practitioners providing landscape architectural services in China. On the contrary, their work faced the challenges of extra costs and uncertain payment. One possible reason that nearly half of the responding Australian landscape architectural firms practiced internationally and more than three quarters of the firms practicing internationally practiced in China was to develop their professional skills and upgrade their professional achievement through engaging more fully in contemporary environmental issues. Such environmental issues are now typified by and accumulated in the urban development in Asia generally and in China in particular. This is evident in the rapid growth of Australian landscape architectural practices in China's urban domain, where most of the work was identified as being carried out in urban public space and a sheer increase of services was being provided for the development of the urban fringe. In addition, the progressive diversification of project types for which Australian practices have been involved in China is an indication of the expanding field for the profession there.

The survey findings also reveal that Australian landscape architectural practice in China appears to be facilitated by the existing professional links that those firms have with other consultancy practices in Australia. In providing services for China, Australians have encountered challenges caused by small firm sizes, despite small-to-medium-sized businesses playing a key role in the economic relations between the two countries. This suggests that small- or medium-sized Australian landscape architectural practices may be providing services to China under the protection of bigger firms or other global providers of services. Such an approach, however, reduces the control of the small-sized firms, which is seen as essential for quality project outcomes.

The survey also showed that Australian practitioners understand the importance of achieving mutual understanding with Chinese clients and having fuller participation by maintaining control over the construction phase. This is also evident in the increasing number of branch offices being established between 1995 and 2005 and the use of them as bases for coordinating construction administration and landscape maintenance. However, during the pre-design services phase, the preliminary stage when mutual understanding with clients is under building, periodic visits and e-communication are still the dominant methods of local engagement. This suggests a trajectory of providing landscape architectural services to China as: first, pre-design phase as less or non-locally based; second, the schematic design phase as a process of progressive localisation; and third, phases of and after construction as localised practice. In short, Australia's provision of landscape architectural services in China was characterised by off-shore planning and on-shore designing practice.

References

Andrews, E. M. (1992). *Australia and China: The ambiguous relationship*. Xiamen: The University of Xiamen.

Australian Bureau of Statistics. (2007). International Trade in Services by Country, by State and by Detailed Services Category, Calendar Year, 2006. Retrieved January 14, 2008, from http://www.abs.gov.au/AUSSTATS/abs@.nsf/DetailsPage/5368.0.55.0042006?OpenDocument. (03 May 2007).

Australian Bureau of Statistics. (2008). 2006 Census Tables. 2006 Census of Population and Housing, (Country of Birth of Person) Retrieved April 09, 2008, from http://www.abs.gov.au/websitedbs/d3310114.nsf/Home/census. (08 February 2008).

Australia Parliament Senate Foreign Affairs Defence and Trade References Committee, & Forshaw, M. (1996). *Australia China relations*. Canberra: Parliament of the Commonwealth of Australia.

Beattle, P. (2005). Bilateral relationship between Queensland and the People's Republic of China.

Fung, E. S. K., & Chen J. (1996). Changing perceptions: The attitudes of the PRC Chinese towards Australia and China, 1989–1996. Queensland, Centre for the Study of Australia-Asia Relations, Faculty of Asian and International Studies, Griffith University.

Kendall, T. D. (2005). *Ways of seeing China: From Yellow Peril to Shangrila*. Fremantle, W.A.: Fremantle Art Centre Press.

McLennan, W. (1995). Selected technical services, Australia 1992–93. Retrieved January 21, 2008, from http://www.abs.gov.au/AUSSTATS/abs@.nsf/productsbyCatalogue/ECA11B1246DA7819CA25722E001BFE4A?OpenDocument.

Nevill, J. (2006). Look both ways. Retrieved October 3, 2007, from http://propertycouncil.gravitymax.com.au/nat/page.asp?622=281749&e_page=17330&=.

Productivity Commission. (2000). Review of legislation regulating the architectural profession, Report no. 13, AusInfo, Canberra.

Wingrove, P. (1999). China and the pacific rim. In R. Benewick & P. Wingrove (Eds.), *China in the 1990s* (pp. 245–255). Basingstoke, Hampshire: Macmillan.

Zhang, Y., & Bull, C. (2008). Landscape architecture as international practice: Trends in Australia and China as a Client. *Urban Planning International, 23*(5), 69–75.

Chapter 5
China as a Site of Demand for International Landscape Architectural Services

Abstract This chapter explores China's demand for landscape architectural services that is created by its rapid and vast urban development and the consequent infusion of these services from overseas. This chapter is comprised of four major sections. Section 5.1 explains China's development, which has been promoted by reform policies that were established in 1978, and the subsequent factors bringing international design practices in China: urbanisation, socio-cultural change and new planning and design market conditions. Section 5.2 analyses the trajectories of three major types of international design practices in China: international architectural practice, international planning practice, and international landscape architectural practice. Section 5.3 discusses the debate in the Chinese Society of Landscape Architecture (CHSLA) that took place around the year 2000, which was stirred by the nine-year delay of institutionalisation of landscape architects。 The section also addressed the comparisons made by CHSLA between China and the West of the role of planning in landscape architecture, and of ways to sustain the characteristics of place. Section 5.4 analyses the repercussions of international design practice in China through a review of the discourse of both international and local designers, as well as scholars from inside and outside of China.

Keywords China's urbanisation · Open-door policy · Residential landscape deign · Environmentally sensitive planning · Institutionaliastion

5.1 Forces Bringing International Planning and Design Practice

The infusion of international design practice into China is underpinned by China's rapid development, which was initiated in late 1978. In that year, China established reform policies and consolidated the power of Deng Xiaoping—the president of China during that time. Deng had long advocated and pushed for reform and modernisation in China. Deng's idea of modernisation meant more industrialisation, more rapid economic development, a higher standard of living for as many people as possible, and a transformation of Chinese society into one that is more open to the

Y. Zhang, *Place Making in International Practice of Landscape Architecture*, Advanced Topics in Science and Technology in China 64,
https://doi.org/10.1007/978-981-16-2442-1_5

outside world (Mackerras et al. 1994). Since then, China has achieved impressive rates of economic growth and urban development, and has improved its foreign relations. Benewick and Wingrove (1999, p. 2) described how, in the 1990s, China received significant increases in fixed capital investment and experienced rising standards of living. Foot (1999, pp. 235, 244) also described how China's policies during the 1990s were likely to resemble those of the mid-to-late 1980s, the most productive era in the history of China's foreign relations. Among the developments listed above, international design practice has been driven specifically by China's rapid urbanisation, the changes in culture and way of life, and the new market conditions for planning and design services.

5.1.1 Dimensions of China's Urbanisation

China's urban development has been concentrated in the coastal area, the area in which economic growth has been the most rapid, foreign businesses have been most active, and in which international design practice initially found its foothold. In 1980, China established four Special Economic Zones for four cities in southern coastal provinces: Shenzhen, Zhuhai, Shantou and Xiamen. These cities encouraged new types of development and offered privileges to foreign investors (Mackerras et al. 1994, pp. 90–101). The cities, especially Shenzhen, are near Hong Kong, a territory that was originally leased by the British and that was one of the major sources from which foreign influences came to China. In the late 1980s and later, the Special Economic Zones model was followed by the development of eastern coastal cities, forming an urban developmental zone along the south-eastern coast of China. The rapid economic growth, urban construction and infusion of foreign investment along this coast brought international practice in planning and design. It was here that building design started in capital cities, and since then, it has expanded to the design of open spaces and the planning of urbanising fringes. Foreign design firms have also consolidated their business in China. These firms initially established branch offices in capital cities along this coastal urban development zone, such as Shenzhen, Shanghai and Beijing. They then gradually developed branch offices in secondary cities along this zone, such as Suzhou and Dalian. Through these branch offices, foreign design firms have also expanded working territories to the in-land area of China.

China's rapid urbanisation implies diverse changes that potentially influence planning and design practices. In his book, *China's Urban Transition*, Friedmann (2005, pp. 36–38) explained such diverse change by discussing five dimensions of urbanisation in China's countryside. One dimension is administrative urbanisation, referring to the increase in the urban population that comes with the reclassification of non-urban places to urban places (Friedmann 2005, p. 36). Such administration of the urban population, as explained by Friedmann (2005, p. 36), is a characteristic of China because the Chinese government controls the urban population by distributing residency permits that are conferred with certain entitlements such as access to housing and subsidised foods. According to Friedmann, about a fifth of the increase in China's urban population has occurred as a result of this approach. The second dimension is economic urbanisation, referring to the proportional decline in the primary sector (agricultural, extractive) and the increases in secondary (manufacturing industry and construction) and tertiary (trades and services) sectors (Friedmann 2005, p. 37). The third dimension is physical urbanisation, referring to the demand for improvements in the urban environment, such as the paving of streets, beautification of public spaces and development of high-rising housing, despite the accompanying environmental degradation that is so prominent in China (Friedmann 2005, p. 37). The fourth dimension is socio-cultural urbanisation, which has manifested in the changing of everyday lives, tensions between newcomers and local communities, and the new forms of individual and household consumption and leisure (Friedmann 2005, p. 37). The fifth dimension is political urbanisation, referring to the decentralization of decision-making power to local authorities, including the formation of stronger links between local officials and business elites in power sharing (Friedmann 2005, pp. 37–38).

The five dimensions of urbanization can be further related to the international demand for planning and design practice. Administrative urbanisation leads to rapid changes in land use, because the process of change is controlled by the government through an administrative approach. This can explain the rapid cleaning up of lands that were observed during urban transformation in China and urban construction on a *tabula rasa*. Economic urbanisation suggests an emphasis on economic and technological development and the transformation of rural and productive lands for industrial and commercial development (such as the building of factories, Central Business Districts and tourism development). Physical urbanisation suggests that planning and design are in demand for building urban infrastructure and urban renovation. Socio-cultural urbanisation suggests changes to culture and society that lead to requests for new types of cities, buildings, streets, open spaces and parks. Political urbanisation suggests the presence of diversified stakeholders in urban governance and consequent changes in project delivery and the process of planning and design.

This multi-dimensional urbanisation therefore suggests two dimensions of China's demand for international design practice, including landscape architectural practice. On the one hand, international design practice has been required because of China's lack of planning and design services to accommodate rapid and vast land change and to build urban entities. This dimension is quantitative. On the other hand, it has also been required in the insufficiency of new forms of buildings and spaces, and new ways of planning and design practice. This dimension is qualitative, suggesting characteristics of international design practice in China. The next section further discusses this second dimension of China's demand for international design practice.

5.1.2 *Western Influences and Socio-Cultural Change in China*

Of the many forces of change influencing culture and society in China, the infusion of products, living styles and ideas from the West since China's opening to the outside world in 1978 is the most significant. Lin (1994, p. 92) argued that since the reforms of late 1978, and especially in the mid-1980s, significant cultural changes have been promoted by 'the relaxation of political and economic control, and the existence of more open-minded mass media' in China. Such cultural changes manifest in the whole living environment, including 'a broader and rapidly growing emphasis on convenience and comfort, and on expanding personal mobility and freedom' (Lin 1994, p. 92).

The above pursuit of convenience, comfort and freedom, however had previously been forbidden in the Maoist era (1949–1976). During the 1960s in China, any contact with a foreign country was deemed a serious crime (Xue 2005, p. 48). Xue (2005, p. 31) indicated that from the 1960s to 1977, China cut its political and economic links with the rest of the world to the extent that the Chinese word for "foreign" became equivalent to the word "crime". Western styles were therefore regarded as the enemy of socialism. However, after China's implementation of the open-door policy in 1978, anything from abroad was deemed innovative and fashionable. Such changes in the whole living environment since the mid-1980s in China also suggest influences from Western living styles on socio-cultural urbanisation, the fourth dimension suggested by Friedmann (2005). It also explains China's welcoming attitude toward international design practices, particularly those from developed Western nations.

These Western influences are also manifested in China's emphasis on economic urbanisation, the third dimension suggested by Friedmann (2005). In his book, *Globalization and cultural trends in China*, Liu (2004) investigated cultural changes that accompanied China's economic development in the most recent two decades and its enhanced relations with the outside world. He claimed that cultural change in China has been a 'full-scale integration into globalisation' (Liu 2004, p. 1). Yet, in Liu's view, China's recent development, especially after 1995, has emphasised technological and economic sectors, rather than the cultural sector. Liu (2004, pp. 1–2) pointed out that:

> These emergent formations, transformations, and reformations cannot be understood merely as the result of China's unique experiments,...Instead, they must be seen within the broad context of globalization.
>
> As an essential area of globalization, culture nevertheless remains peripheral in China's debates and policies, which focus primarily on economic sectors. Similarly, China's studies on the West concentrate largely on the political and economic consequences of globalization.

Liu further explained the decline of cultural studies in China after 1995 and the favour placed on scientific study in the economy, politics, and society, saying (Liu 2004, p. 70) that:

> In the 1980s [the decade of cultural fever], debates and controversies were mainly carried out in the humanities—philosophy, history, and literature—and were focused on China's tradition and modernity in highly eclectic ways…In the second half of the 1990s, a significant shift occurred. The "hard-core" social science[s] entered the stage and soon the discussions of economists, political scientists, [and] sociologists dominate[d] the scene.

Liu's argument suggests that in China's connections with the outside world, cross-cultural studies remained not focusing on the humanities. It could be argued that the invitation to international design firms was thus informed by cross-cultural studies on the economic and political sectors, rather than by the cross-cultural study of the consequent change of people, community and their life.

Western influences on China's living environment in the 1980s and the government focuses on the economy and technological development in the 1990s suggest three stages of China's construction of urban areas that are also manifested in international design practice. In the 1980s, China's rapid urban development was initiated. The goal was to increase the amount of urban entities, i.e., the production of more housing and schools. It was a stage of physical urbanisation during which more buildings were erected. As a result, cities grew denser and upwards. This stage also took place during the decade in which cultural discourse regarding Chinese traditions and modernity became feverish and the way of life of the Chinese people was influenced perceptibly by the West. These two trends, the physical construction process and cultural eclecticism, however, were not connected. During this period, on the one hand, international styles of architecture flooded into China, just as described by Xue (2005, p. 32): 'to most Chinese throughout the 1980s, anything foreign was always thought to be inherently superior', and on the other hand, local architects made efforts to use traditional architectural forms for contemporary buildings, but often merely for decoration. From the perspective of the characteristics of China's demand for international design practice, this has generally been to follow western fashions.

In the 1990s, there was increasing interest in the technological and economical sectors of the West. The construction of urban areas consequently focused on economic urbanisation: rapid changes in land use for industrial and commercial development at the expense of the natural environment. Cities were not only growing denser, but also bigger. It was during this period that international planning practices expanded the working territory from architectural practice to planning new cities. During this period, China's demand for international planning and design services extended to Western political and economical systems, shifting from building styles to urban forms.

During the 2000s, changes in the ways of life and leisure are increasingly recognized in China, as well as the impact of rapid development on society and tradition, and on environmental degradation. During this period, the physical construction has been connected to cultural eclecticism in China. Therefore, China's demand for international design practice has become more culturally and environmentally sensitive during the most recent decades. It was also in this period that international practice in landscape architecture has extended the working territory from residential landscapes to the wider range of the public domain.

5.1.3 New Planning and Design Market Conditions

The infusion of international planning and design services has been promoted by the new market conditions in China. This reflects the fourth dimension—the political urbanisation that brought changes to power-sharing and decision-making in urban development, and changes to the relations between local planners and designers and the demand side of urban construction. The latter was shown in the emerging business qualities of planning and design practice. Early in 1980, the Fifth Congress of the Architectural Society of China made the decision to allow design institutes to charge design fees and to contract with clients (Xue 2005, p. 171). This policy was different from that of the planned economy of the previous time, when designers were paid fixed salaries, regardless of the quantity and quality of their work.

In 1994, at the National People's Congress, a socialist market economy was fully recognised in China (Benewick and Wingrove 1999, pp. 273–274). This means that in the mid-1990s, market conditions prevailed for local planning and design institutes whose work had previously been guaranteed by the government. It was also during this period that China showed a focus on economic and technological development. In this shift, planning and design institutions started to transform their administrative structures and ways of providing services. Meanwhile, such transformation created opportunities for international design practices, which had been used to service market economies in their own societies, to receive work from China. Wu et al. (2007, pp. 159–160) suggest that planners and designers in China have suffered a loss of confidence since the mid-1980s because of the shock-waves that were generated throughout the political and economic arena by market-based reform and globalisation. Hence, local planners, designers, and clients have been willing to import Western theories and consultancy practices.

A freer, more competitive business environment has been in place during the post-reform era, especially after China's entry into the World Trade Organisation in 2001. This further attracted foreign design firms to China to compete with local planning and design institutes. In this post-reform era, international practice in landscape architecture has flourished and expanded its working territory. The following section discusses three types of international design practice in China—architectural, planning and landscape architectural practices, where landscape architects are involved.

5.2 The Trajectory of International Planning and Design Practice in China

International design practice in China began with architecture in the late 1970s and early 1980s. The first foreign influences came from expatriate Chinese architects, such as the American-Chinese architect I.M. Pei and his design of the Fragrant Hill Hotel in Beijing. They also came from ethnically Chinese areas, such as Hong Kong

(Xue 2005, pp. 31–32). Since the 1990s, international design practice in China has come from all over the world and permeated every sector, from building design to planning and designing new cities, urban precincts, and open spaces. Ruano (1999) suggested that international practitioners in China appear as strangers in paradise and have contributed to the instant urban transformation of China. Ruano (1999, p. 30) pointed out:

> Foreign architects are not only brought into the country by foreign investors looking for reliable and competent firms which will not need to go through a learning curve on their projects: Chinese companies (including some owned by public bodies) are also increasingly calling in foreign expertise to provide their projects with the latest architectural fads, state-of-the art construction technologies, project-management capabilities, and no less important, with a famous signature.

5.2.1 International Architectural Practice: Starting with the Spectacular

International practice in architecture started in China with the design of spectacular buildings, initially in regional capitals such as Beijing and Shanghai. The study of *Projects Designed by Foreign Architects in Beijing since 1978* (Li 2005) shows that, during the period from 1982 to 1987, 70% of projects were hotels. That figure decreased to 34% over the next five years, as the number of building complex projects increased from zero to 42%. From 1994 to 1998, hotel projects dropped to 5%, and building complex projects increased to 52%. "Building complex" refers to buildings with comprehensive functions, such as those of shops, offices and other functional centres. The more recent major participation in designing building complex suggests the relevance of international architectural practice to urban design and planning (Fig. 5.1).

Since the 1990s, international design practice has spread in China and in diversified territories. First, many major cities in China achieved significant economic growth and were able to afford the participation of overseas architects for prominent and technically complicated projects (Xue 2005, p. 36). Second, the variety of international architectural practice expanded gradually to government office buildings, schools, health care facilities, residential buildings, cultural facilities, and landscaping and urban planning, even to urban renewal and conservation (Xue 2005, p. 36). It could be argued that the expansion of working territory from the design of buildings to other areas was promoted by the initial international architectural practice of designing spectacular buildings. The central position of these buildings in cities, such as the previously mentioned building complex, and their attached public functions such as the provision of transportation nodes and cultural and knowledge centres (Ruano 1999, p. 29) promoted the expansion of international architectural practice to urban design and planning.

The expansion of international architectural practice in China in turn promoted the infusion of foreign architectural firms, both renowned and less famous (Xue 2005,

Fig. 5.1 International architectural practice—spectacular buildings. *Source Projects Designed by Foreign Architects in Beijing since 1978* (Li 2005, p. 17)

p. 36). The study of projects designed by foreign architects in Beijing also shows that the participation of architectural firms from Europe increased in the late 1990s at the expense of American and Japanese architectural firms (Li 2005, p. 18). The more renowned architectural firms continue to participate in the design of spectacular buildings, and are consequently involved in urban design and planning at present in the 2000s. Less famous firms are engaged in a variety of medium and small-scale projects, especially projects of residential sectors. In the article, *the Intergrowth of Chinese Architecture: the foreign and the local*, Lou (2005) cited figures provided by a survey of real estate advertisements in Beijing. The survey shows that, of the twenty-eight advertisements for new developments during the period from the 18 March to 20 April 2005, fourteen were designed by foreign firms and twelve were designed jointly by foreign and local firms, but only two were designed solely by local firms. The extension of international architectural practice in urban design and planning and the broader participation in residential projects in China suggest the involvement of international landscape architectural work and landscape architects.

5.2.2 International Planning Practice: The Conceptual and Design-Based

International planning practice in China was initiated by the international architectural practice of designing spectacular buildings, as argued previously. However, its spread has been underpinned by the shift of city planning from the allocation of resources to the promotion of places. Wu et al. (2007, p. 180) described this shift as 'the intense intercity competition and promotion of place-based strategy in order to attract mobile capital and sustain local growth'. In this process, international planning practice has taken the forms of conceptual and design-based planning, which are two of the four new planning practices emerging in China. Wu et al (2007, pp. 180, 182) argued that these four planning practices emerged to deal with the new urban reality that current urban planning systems can hardly fulfil. It could be argued that the conceptual and design based planning practice tend to beautify the urban landscape.

International conceptual planning practice means foreign participation at the strategic level of urban growth, because overseas planning firms are not allowed to conduct master planning for Chinese cities. Wu et al (2007, p. 180) explained the typical process of inviting foreign firms:

> A common practice is that the local government first invites a foreign firm to devise a concept plan, and then asks a qualified domestic design institute to incorporate the ideas of the foreign firm into a master plan that will be submitted to the upper level for final authorization. This practice reflects the aggressive pursuit of new urban strategies in Chinese cities. In order to gain international recognition quickly, city planning projects are often branded by the world's top architects and planners, and local government then make full use of their 'signatures' to achieve broader urban significance...Finally, the concept plan may form a material consideration in development control but might not be accorded equal weight with statutory plans, such as the master plan and zoning.

International design-based planning in fact means urban design. Wu et al. (2007, p. 182) regarded this form of planning practice as 'an art of making places' and 'involves the design of buildings, groups of buildings, spaces, and landscapes'. They (Wu et al. 2007, p. 183) pointed out:

> Similar to the concept[ual] plan[ning], localities often invite world-renowned architects and planners to undertake large urban design projects...Over the past decades, renowned planning institutions, such as Kisho Kurokawa and Arata Isozaki from Japan, SOM from the US, Atkins from the UK, and GMP from Germany, have left their names on a number of landmark projects through design competitions and bidding...It is hoped that by opening the urban design market the cities will soon get close to the world standard of urban construction.

The above discussion suggests a characteristic of the international planning practices in China that is an absence of laws and regulations. Wu et al. (2007, pp. 180–183) also pointed out that international conceptual planning can overcome the inflexibilities of statutory planning, and urban design is a relatively new practice, so operational guidelines have not yet been formulated by the central government of China. Figure 5.2 shows the concept planning by architects Von Gerkan, Marg and Partners in 2002 for the Ling-gang New City of Shanghai, and the urban design by Bing Thom Architects in 1994 for the city of Dalian. All of these suggest that world-class architects have been key players of international planning practices in China.

Two further new planning practices emerging in China are environmentally- and culturally-sensitive planning. These two planning practices, however, were not argued as being associated with international planning practice by Wu et al. (2007). In contrast, they use international architectural practice as negative examples in arguing for culturally sensitive planning in China.

Environmental sensitive planning is characterized by administrative strategies for the control of draft planning. Wu et al. (2007, p. 186) argued that this type of

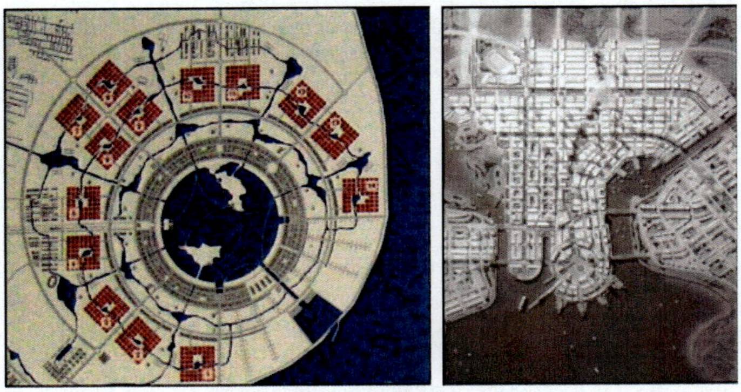

Fig. 5.2 International planning practice—the conceptual/design based. *Source* Left—*Projects in China: architects Von Gerkan, Marg and Partners* (Gerkan 2008, p. 30); right—*Urban development in post-reform China: state, market, and space* (Wu et al. 2007, p. 184)

new planning practice is guided by policy initiatives and administrative tactics for the purposes of reinforcing more responsible local attitudes toward environmental protection. This environmentally sensitive planning is also monitored by different Environment Impact Assessment (EIA) requirements set out by law for two categories of government plans. One category is sectoral plans. The other is guidance plans for land utilisation, and the development of watershed areas and sea areas (Wu et al. 2007, p. 186). The policy initiatives also include setting up environmentally sensitive areas, including regions of special landscape, wildlife, or historic interest, which are considered vulnerable to the intensification of urban development (Wu et al. 2007, p. 186). Wu, Xu et al. provide an example of environmentally sensitive planning for Beijing's outer greenbelt that sets important green areas for short-term period (Fig. 5.3a) and sets areas under restricted construction (Fig. 5.3b). It could be argued that this emerging, environmentally sensitive planning is particularly relevant to landscape architecture as a modern profession in China.

Culturally sensitive planning concerns heritage, tradition and post-modern influences. Wu et al. (2007, pp. 188–189) argued that cultural and heritage conservation has not yet received sufficient attention in most Chinese cities. In this regard, cities are simply mega projects that are 'branded by the world's top architects and planners', and the space is therefore full of post-modern architecture and urban forms. The authors (Wu et al. 2007, p. 189) indicated that the single-minded pursuit of economic return and wealth creation, together with the lack of a comprehensive cultural policy framework, has created cultural uniformity at the expense of the distinctive qualities of Chinese cities. The authors (Wu et al. 2007, p. 189) cited comments from an American planner who has worked in Beijing for ten years and said that 'some foreign architects whose juvenile designs would otherwise be rejected in their home countries have found sympathetic audience' in Beijing'. Such comments, however, suggest the problem arises in relation to the cultural construction of the urban area that exists both in local expectations and international design practice.

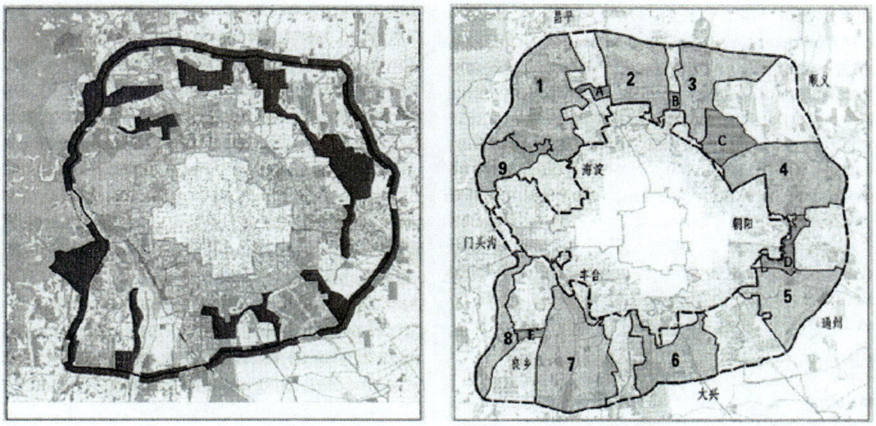

(a) (b)

Fig. 5.3 Local environmentally sensitive planning. *Source Urban development in post-reform China: state, market, and space* (Wu et al. 2007, p. 187)

According to Wu et al. (2007, p. 190), culturally sensitive planning has been emerging in local planning practice, in the supervision by local planning authorities of the preservation of historical buildings and in the emerging awareness among planners of the development of a sense of history, not just by the preservation of historical buildings, but also by the sensitive incorporation of factors unique to each city in their planning and design (Wu et al. 2007, p. 190). Figure 5.4 shows an example of culturally sensitive planning practice—the preservation and renewal scheme for central Beijing, which emphasized the preservation of heritage buildings, their forms and details, in short, the material evidence of the local history. Wu et al. (2007, pp. 190–191) indicated that in addition to the conservation and revitalisation of individual buildings, few cities in China are really achieving a culturally sensitive planning practice that can sensitively incorporate the uniqueness of place. The lack of success in culturally sensitive local planning practice suggests greater challenges to international culturally sensitive planning practice.

5.2.3 International Landscape Architectural Practice: The Recognised

A clearer recognition of international landscape architectural practice in China started at the turning point of the twentieth century, through landscape architectural firms' broader participation in projects in residential sectors and along waterfront areas. Before that, international landscape architectural practice was concealed under international architectural practice delivered by renowned foreign architectural firms, as

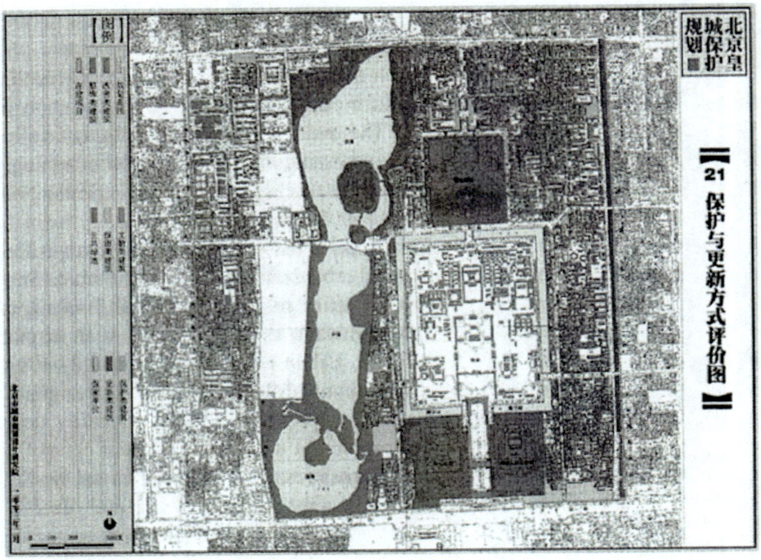

Fig. 5.4 Local culturally sensitive planning. *Source Urban development in post-reform China: state, market, and space* (Wu et al. 2007, p. 190)

analyzed previously. This clear recognition is evident in the discussion of these practices in core professional journals in China, in particular, the journal of Chinese Landscape Architecture (CLA).

International landscape architectural practice in residential landscape design was initially provided by Hong Kong-based foreign landscape architectural firms. In his article, Ye (2003) described the advantageous position of these firms in China's design market. He cited a survey taken in 2001 of the landscape designs submitted for ten model residential estates in Shenzhen, a city bordered by Hong Kong. The survey found that of the ten model estates, the outdoor spaces of eight were designed by foreign landscape architectural firms, including three by the Belt Collins Hong Kong office (Ye 2003). Similarly, Lin (2003) described the flourishing foreign landscape architectural practices in China by discussing the work provided by this Belt Collins Hong Kong office. Lin described the firm as 'a warmly welcomed company in residential landscape design in China', saying that eighty percent of the business of the Belt Collins Hong Kong office came from residential projects located in mainland China. Moreover, the business of the Belt Collins Hong Kong office had exceeded that of its head office in the US. Lin (2003, pp. 47–48) asserted that the success of Belt Collins in the Chinese market arises from three characteristics of its practice. One is the application of what he described as the 'professional design of water feature and its modern American styles'. The second is the maintenance of good relationships with clients, based on its strategy for business operations and experience in planning and design over fifty years. The third is their strategies for quality control, including the selection of landscape materials and supervision of construction.

Such reasons as those listed by Lin suggest the demand in China for foreign landscape architectural services included an emphasis on the use of water features and project management. Published in 2003 by the principal professional journal CLA, Lin's appraisal of the use of modern American styles of water features in contemporary residential landscape in China suggests a trend in the attitude of Chinese local practitioners towards looking for new forms from the West to accommodate new ways of life at that time.

In addition to the typical discussion of international landscape architectural practice by local Chinese design practitioners and scholars, the American firm EDAW in 2004 became the first foreign firm to introduce its projects to China through the journal CLA. Lau (2004, p. 15), the regional director of EDAW in Asia at that time, described the development of EDAW in the design market of China, saying that EDAW's practices in China began with residential projects and developed into comprehensive planning, and urban and public space design. Of the projects submitted by Lau to CLA, five were selected and published by CLA, of which all were waterfront projects. These projects include 'Jin-ji Lake' in Suzhou, started in 1998; 'Five Mile Lake [Li Lake]' in Wuxi, started in 2002; 'Suzhou Creek' in Shanghai, started in 2002; 'Lu-chao New Harbour City' (or Lin-gang new city) in Shanghai, started in 2003, and '2008 Olympic Aquatic Park' in Beijing, started at the end of 2003 (Lau 2004, pp. 16–21). Lau (2004: 15) explained that the success of EDAW can be attributed to their integrated approach, saying:

> Our integrated approach that balances market driven forces with concerns for the environment has been increasingly well received in China and Southeast Asia, where the governments and development communities are quickly learning the value of proper planning in realizing the highest and best value of land and water resources.

By 'integrated approach', Lau (2004, p. 15) referred to EDAW's trademark "DEEP" planning approach, which stands for the integration of Design, Environment, Economics and Planning.

These waterfront projects introduced by the principal professional journal CLA also suggest that the attention of the Chinese Society of Landscape Architecture (CHSLA) is focused on lands bordered by lakes and rivers and the consequent demand for foreign landscape architectural services for their development. It also suggests foreign landscape architectural firms' alliance with local governmental sectors in China, rather than with the private sector alone, as had previously dominated work on residential estates. Such development in the working territories of international practice in landscape architecture in China also suggests a form of environmentally sensitive practice. The expansion of international architectural practice to ecologically targeted new cities in the late 1990s, such as international tourist towns and villages (Ruano 1999), also suggest the transformation of architect-led planning practices to landscape architect-led environmentally sensitive planning.

International landscape architectural practice has also been recognized in the design of urban parks. In the journal CLA, Wang and Liu (2007) described a park jointly designed by the Shanghai Landscape Architectural Design Institute and the Canadian firm, Williams, Asselin, Ackaoui & Associates (WAA) in Shanghai. Wang and Liu (2007), who are from the Shanghai Landscape Architectural Design Institute, described how the design incorporated the history of the site and the everyday lives of the public (Fig. 5.5). Wang and Liu (2007, p. 43) claimed that the success of the park design is based on the design method that 'interprets memories of Shanghai local residents' and 'tells the ever-happening stories of Shanghai'. Although the role of WAA was not explained in that article, WAA (2002) claimed on its website that the Xu-jia-hui park is successful in its 'reflection of the regional characteristics of Shanghai'. In the constructed park, the history of the area is recalled through the laying out of paths that connect heritage buildings and artefacts. New forms of landscape, modern structures and materials are introduced, while rocks and bamboos are positioned in a way that suggests traditional Chinese landscape gardening (Fig. 5.5). The design of paths to recall local memories and stories suggest concern for the non-material history in this local-international jointly designed projects.

The flourishing of foreign landscape architectural practice in China is also evidenced in the reflections of local Chinese design practitioners on their own positions. Zhang (2006, p. 60) argued that global interconnectedness reserves the frontiers of landscape architectural practice in China for Western landscape architects. Meanwhile, local Chinese designers are often preoccupied with learning Western theories (Zhang 2006, p. 60). Zhang (2006, p. 60) pointed out that under such conditions, China has become an experimental field in which to test Western thoughts and

Fig. 5.5 International landscape architectural practice—Xu-jia-hui Park. *Source* Author, photographed in 2003

theories. She (Zhang 2006, p. 61) claimed that landscape architectural practice in contemporary China is consequently suffering amnesia, which means that the many designed and constructed landscapes in present-day China suggest the disconnection from China's history and tradition.

The infusion and flourishing of international landscape architectural practice in China at the end of twentieth and beginning of the twenty-first centuries have been greatly influenced by the changing status of the local profession of landscape architecture in China. The following section further analyses such a status and explores the consequent demand for international landscape architectural services in China.

5.3 Stirrings in Landscape Architectural Discourse in China: 1995–2006

The broader infusion of international landscape architectural practice in China since the late 1990s has been influenced specifically by the nine-year delay of the establishment of a registration system for landscape architects in China. Stirrings in the CHSLA during 1995–2006 coincided with the expansion of international landscape architectural practice from the residential landscape design to environmentally sensitive planning. Two topics are identified from this theoretical debate. One is the role of planning in landscape architecture; the other is approaches to a sense of place. Both are discussed in comparison with the modern West and for the purposes of continuing the tradition of Chinese *Yuan-lin.*[1]

5.3.1 The Nine-Year Delay in Institutionalisation

In 1995, a proposal for a landscape architects registration system was considered by the National Construction Ministry of China[2] as part of its overall work plan (Meng and Wang 2002; Wu 2004). One intention in establishing the professional registration system was for China to align with international standards, thereby preparing for membership of the World Trade Organization (WTO) (Wu 2004). It was also expected that registration systems would ensure the equality of local designers within an increasingly internationalized design market in China (Wu 2004). Although the work plan eventually nominated 1998 as the year when a landscape architect registration system would be enacted, it was in fact a further eight years (2006) before this occurred, the year that the CHSLA joined the International Federation of Landscape Architects (IFLA), the body representing landscape architects worldwide.

---

[1]Yuan-lin is interpreted as 'gardens; park' in Wei (2006).

[2]In July 2008, the National Construction Ministry of China was restructured as a part of the National Ministry of Housing and urban-rural development.

The 1995 proposal to establish a registration system for landscape architects suggests the level of demand for landscape architectural services in China. Yet, the delay in institutionalisation for almost a decade resulted in uncertainty for local Chinese landscape architectural practitioners. Conversely, it left many opportunities for international landscape architectural practitioners. One reason for this delay, as was explained by Ru-tang Ye, the undersecretary of the National Construction Ministry at that time, was the small number of landscape architects then practicing in China (Wu 2004). However, the professional debate in CHSLA shows a range of work relevant to landscape architecture since the early 1950s. The debate also shows the problem of the widely held belief in Chinese society regarding landscape architecture as merely landscape gardening or providing green space. During this nine-year period, the registration system for architects was established in 1995. The registration system for urban planners was established in 1998. China finally entered the WTO at the end of 2001. China's entry into the WTO further freed up the market for international planning and design practice. All of these resulted in greater opportunities for international landscape architectural practice in China.

The delay of the institutionalisation of landscape architecture as a modern profession also took place in the establishment of professional journals. First published in 1985, the journal *Chinese Landscape Architecture* (CLA) is relatively new in China, especially in comparison with *Architecture Journal* and *Urban Planning*, which were first published in 1954 and 1977, respectively. The delay in developing professional journals again reflects the shorter history of landscape architecture as a distinctive modern form of planning and design practice in China.

The period between 1995 and 2006 was therefore full of controversial questions and debates regarding the values of Chinese garden traditions, the role of contemporary landscape architecture in China's urbanisation, and the urgent matter of solving environmental degradation. International landscape architectural practice in China, especially during the boom, further stirred up the comparisons by CHSLA with the West regarding their theories and practices.

The following paragraphs analyze two central topics: the role of planning and approaches to a sense of place in landscape architecture. The analysis is based upon articles selected from the abovementioned journals, in particular CLA, whose topics on foreign landscape architectural practices in China surged in 2000. Most of the articles selected were written by local Chinese landscape architectural practitioners in the Chinese language and cited in this research in English. Some of the writers are design scholars in China. These articles also include several written by Western practitioners and observers that have been translated and published in these professional journals in the Chinese language.

5.3.2 The Role of Planning: Comparisons with the West

Comparisons of the role of planning in Chinese landscape architecture with that of the West were evident in the CHSLA debate regarding the Western term 'landscape architecture' and its counterpart in the Chinese language. In 2004, there was intensive argument centring on whether the long-used Chinese term *Yuan-lin* (gardens and parks) is sufficient to explain the modern meaning implied by the Western term 'landscape architecture'. A central point in the debate was the role of planning. In relation to large scale land change, Liu et al. (2004, p. 42) pointed out that the education and practice of landscape planning was not yet launched in China compared with the developed West. This explains in part the demand for international landscape architectural practice in planning.

Similar to the views of Liu et al. (2004) regarding the role of planning in landscape architecture in China, Yu (2004, p. 38) indicated that the scope of planning in China falls short of international standards. In comparing with planning practice in the US, Yu (2004) argued that the professional scope of landscape architecture in China was limited to the design of urban green spaces, whereas in the US, new city planning is a major task for landscape architectural practice. Yu (2004, p. 37) described how some American landscape architectural firms and individual landscape architects are renowned for their expertise in planning new cities. Further, Yu (2004, p. 38) argued that the understanding of landscape architecture in China was then limited to landscape gardening, and that this aestheticism of understanding is harmful, reducing China's capacity to address the overwhelming process of urbanisation and dramatic land change across its vast territory. Yu (2004, p. 38) asserted that such limited aesthetic purpose for landscape architectural practice is a problem of 'the whole Chinese society'. Although extreme, Yu's argument exposes a deep-rooted and common understanding of landscape architectural practice in Chinese society, especially in the understanding of non-professionals, which relates to the inclination towards a picturesque landscape. It could be argued that this inclination is related to the rooted Chinese ways of thinking that is identified as 'aesthetic' (Hall and Ames 1995, p. 6), as discussed in Sect. 2.3.2. Later, Yu (2006, p. 73) also called for new ways of creating a vernacular landscape, saying:

> Facing the environmental reality in present-day China and scrutinized by environmental ethics, the traditional Chinese principles, the so-called "man and heaven are unified" and "let nature take its course", seem dishonest and sound empty.

The debate stirred by the contrasting roles of tradition with those of planning in landscape architecture in China presents two divergent attitudes towards landscape architectural practice in Chinese society. One emphasizes the preciousness of Chinese garden tradition and the underpinning literary culture, looking to the past. The other emphasizes the urgent need for the use of environmental strategies and its underpinning modern technology, looking to the future and the modern West. Of those emphasizing environmental strategies, Yu (2004, p. 38), for example, claimed: 'contemporary landscape has been assessed by scientific standards and advanced technologies of ecology, biological diversity and environmental protection, rather than forms'. These two divergent attitudes towards practice are further embodied in the discussion of ways to respond to sense of place in China, again comparing the principles and projects of the West with those of China.

5.3.3 *Sustaining the Characteristics of Place: Comparisons with the West*

Comparing approaches to sense of place with those applied in the West suggest the attempts of CHSLA to search for common ground between the traditions of Chinese *Yuan-lin* and the modern meaning of the Western term "Landscape Architecture". Yu (2000) argued that China and Western countries have a shared experience in their search for lost place. Firstly, all experience how place becomes meaningful through living unconsciously there. Secondly, all reflect upon the loss of place caused by industrial and urban development. Finally, all have made efforts to use design as a conscious means of regaining the sense of place. Yu outlined three landscape architectural approaches to maintaining and developing a sense of place that can be shared across cultures. These include first, the pre-scientific approach that is exemplified by the Chinese term '*feng-shui*', which translates as 'wind-water' and relates to geomantic qualities (Yu 2000, pp. 3, 45). The second is the phenomenological approach, exemplified by Lynch's 'The Image of the City', and the 'Genius Loci' of ancient Rome. The third is the ecological approach, exemplified by McHarg's 'Design with Nature' and his planning strategies, and the subsequent 'landscape ecology' (Yu 2000, p. 46). Yu's comparative studies can be further related to his 2006 speech at the 43rd IFLA world congress in the US, in which he claimed that the landscape architecture of China is inevitably 'of the world' (Chen and Ge 2006). Such a claim suggests that ways to mediate environmental problems currently manifested in China and to sustain the natural and cultural characteristics of places are valuable to the world.

Similarly, in Zhang's (2006) comparative study of traditional Chinese and modern Western design philosophies, sense of place is again raised as a key concept and compared across cultures and times. Zhang (2006, pp. 51–52) argued that the concerns regarding human spirit in landscape architectural design are always centred on the meaning and experience of place. Regarding China, Zhang suggested that the fundamental impacts of the West on landscape architecture come from Western design philosophies. Compared to the 'functionalist', 'people-centred' and 'nature-oriented' design philosophies of the West, Zhang argued that the Chinese traditional philosophies of 'man and heaven are unified' (*tian-ren he-yi*) and 'following nature' (*dao-fa zi-ran*) are similar to the latest trends in the West (Zhang 2006, p. 52), (nature-oriented design philosophy). In particular, Zhang referred to the American naturalist Aldo Leopold and his 1949 book, *A Sand County Almanac, and Sketches Here and There*, suggesting the importance of change from the 'market-oriented', 'consumer-oriented' and 'people-centred' design philosophy to traditional Chinese and 'nature-oriented' Western design philosophies. Such Western design philosophies as recommended by Zhang (2006) as 'nature-oriented', however, were first articulated in the West almost fifty years ago and have long antecedents in Western culture.

Furthermore, the practical work of landscape architectural practices in the West and China are compared regarding the way in which they respond to sense of place.

This is evident in the argument of Han and Shen (2005) for sharing 'place' (*chang-suo*) principles between the West and China, as illustrated by landscape architectural projects worldwide. They (Han and Shen 2005, p. 61) claimed that design based on regional features is a principle shared between both Chinese classical gardens and Western modern landscapes. Han and Shen (2005, p. 62) listed climate (*qi-hou*), place (*chang-shuo*), context (*bei-jing*), culture (*wen-hua*) and society (*she-hui*) as five key aspects of regional sense of place. They provided examples of each aspect, such as the use of native plants, respect for local topography, history and culture, and the role of landscape in society. Their Western examples include Thomas Church's Donnell Garden (Fig. 5.6, top left) and Roberto Burle Marx's Brazilian Gardens (Fig. 5.6, middle left), as examples that are responsive to 'climate'; George Hargreaves' Byxbee Park (Fig. 5.6, middle right), designed in 1988, as an example responsive to 'place';

Fig. 5.6 Approaches to a sense of place—Western examples. *Source Chinese Landscape Architecture* (Han and Shen 2005: 62, 63, 66, 67)

Isamu Noguchi's 1983 stone sculpture for the landscape design of California Scenario (Fig. 5.6, top right), as an example responsive to 'culture'; the Duisburg Nord Park (Fig. 5.6, bottom left) in Germany designed by Peter Latz in the 1990s and Suzhou Jing-ji Lake in China designed by EDAW in 1998, as examples responsive to 'society' (Fig. 5.6, bottom right).

Examples of projects in China, however, are centred on recent planning and designs associated with lakes, parklands, urban precincts and open spaces, which were designed by local practitioners in China, mainly after 2003 (Fig. 5.7). Comparing the two sets of examples, those of the West are about ten or twenty years older, suggesting the discrepancy between place examples designed in the West and in China. In addition, a nuance can be identified in the expression of sense of place between the Western and Chinese examples. The former suggests the integration of the context in a subtle way. The latter suggests the focus on the natural characteristics and the conservation of traditional architectural forms.

Fig. 5.7 Approaches to a sense of place—Chinese examples. *Source Chinese Landscape Architecture* (Han and Shen 2005: 64, 65)

More recently, the term 'new regionalism' has been used in landscape architectural discourses in China, promoting the recognition of local characteristics in a global context. In his article, *Applications of new regionalism in urban landscape*, Wang (2007) uses 'new regionalism' to explain how local characteristics are maintained in the face of challenges resulting from changes brought by development and globalisation. He indicated that the approach of new regionalism is a way to respond to genius loci (*chang-suo jing-shen*) in the design of urban landscapes. Compared to regionalism, which proposes design solutions based upon the use of the symbols and patterns of local cultures, Wang (2007, p. 60) argued that new regionalism accepts the dynamics of cultural conditions and concerns with the 'authentic, familiar trajectories' of everyday life, suggesting that they are significant components of a regional sense of place (*di-yu-xing*). Such an argument suggests the increasing concern in CHSLA with the implicit component of a sense of place: the local or original community and their lives.

The above discussion suggests a trend in attitudes in CHSLA towards looking for a shared platform of design principles and philosophies between the modern Western and traditional Chinese approaches. If such a platform can be found, it demonstrates the possibility of using tradition in contemporary practice, which can then serve China's modernisation and the rapid cultural change brought on by globalisation, and solve the perceived loss of place resulting from China's rapid development and Western influences. Yet, such comparisons with the West, focusing on Western theories and practice decades earlier, overlook the fact that present practices and thinking have already been influenced by the cross-cultural conditions of the contemporary global period.

Instead, the claimed similarity between Chinese tradition and philosophy and that of the modern West contrasts with the reality of omnipresent Western forms and the break with tradition in the designed and constructed landscapes of contemporary China, resulting in the perceived loss of place. Liu (2000, p. 27) pointed out that valuable Western concepts such as 'regionalism' (*di-yu zhu-yi*), 'eclecticism' (*li-shi zhu-yi*) and 'contextualism' (*wen-mai zhu-yi*) are often disregarded in the use of Western models in China. Liu (2000) proposed two general and opposing trends in landscape architectural discourse and practice in China. The first is the nostalgic praise of the tradition of Chinese classical gardens. The second is the copying of landscape forms and design methods from the West. Liu (2000) asserted that both trends provide no solutions to the challenges of retaining local characteristics and ensuring cultural continuity. Specifically, Liu (2000) argued that negative influences from the West are looming large in the forms of contemporary parks in China. Similarly, Yan and Lu (2008, p. 16) asserted that contemporary Chinese landscape architecture has been caught in a dilemma with foreign landscape forms reaching the zenith of their influence at the very time that Chinese tradition has found it difficult to continue in present design.

5.4 Repercussions of International Design Practice in China

International landscape architectural practice is seen in international architectural and planning practices in China and covers a wide range of working territory. Yet, the most recognized of international landscape architectural practices in China is the design of residential landscapes. The flourishing of the abovementioned international practice in planning and design in China, including landscape architectural practice, has caused wide reverberations. These reverberations are from planning and design practitioners and scholars, researchers and educators, both inside and outside China, and suggest that challenges and problems are caused by the practice.

In international architectural practice, challenges and problems are evident in the reflection of a few foreign architects who have practiced in China. Ruano (1999, p. 30) indicated that, in addition to the abundant opportunities presented to foreign architects by working in China (especially those from America and Europe), international practitioners have been challenged by idiosyncratic codes, conflicting expectations, communication problems and low design fees.

Similarly, in his first account of his firm's practices in China between 1998 and 2008, the German architect Gerkan (2008, pp. 8–16) from GMP suggested that the architectural practice of his firm in China is mainly challenged by the operational methods of Chinese clients, whereas challenges to their planning practice of new cities are more complicated, as 'the master plan[ing] is one thing, implementing it is another' (Gerkan 2008, p. 15). Gerkan also pointed out the problem of responding to the tradition of Chinese architecture in both international and local architectural practices. Gerkan (2008, p. 11) claimed:

> [The] fact [is] that among the many new buildings erected in recent years in China there are hardly any noticeable differences between the work of Chinese architects and that of foreigners. They all apparently have in common: the imitation of American models, the use of materials that symbolize high-tech and progress and the combination of these with supposedly traditional Chinese elements as decoration.

The above discussions show a fundamental challenge to international design practice caused by Chinese operational procedures. It also shows that the reality of urban construction in China contrasts with the good intentions of both local and international practitioners in considering Chinese traditions.

In international planning practice, similar challenges and concerns exist. Referring to his practice in planning new cities, Gerkan (2008, p. 13) claimed that his practice has had more opportunities to be responsive to Chinese traditions. Reflecting on the urban planning project of *Lin-gang* (Bayside) New City, located sixty kilometres away from Shanghai, Gerkan (2008, pp. 13, 15) described the approaches of his practice to Chinese tradition through 'creating units of reasonable dimensions that are not anonymously absorbed in the urban structure,' and design 'the facades, forms and materials' for a new city 'look like China rather than…[a district of Hamburg, Germany]'. Such a comment, on the one hand, again shows concern for Chinese traditions by international practitioners. On the other hand, it suggests that the architects' vision of sustaining the characteristics of place in planning practice is focused on physical entities and details, rather than responding to the more subtle components, such as lasting memory of a place and way of life.

The problem of expressing a sense of place in planning was further argued by Wu et al. (2007). They (Wu et al. 2007, p. 191) claimed that although western theories and practices have been welcomed along with the exercising of new planning practices in China, the unsolved problems of planning implementation and the expression of place benefit little from those imported from the West:

> Planners in China have begun to exercise new planning practices. Government officials and planners also hope to import new technology from Western countries to modernize urban space. Indeed, new methods and technologies in urban planning are exposed to Chinese planners by increasing international academic exchanges and consultation practice. However, the mere introduction of new technology cannot improve urban planning in China. Effective planning control relies…on the implementation side, on the philosophy of how plans should be made, and whether the newly designed planning is compatible with the society in which it operates. This also raises important questions regarding cultural uniformity and sustainability. Unfortunately, many planning practices do not contribute to a regional expression of place.

Likewise, Friedmann (2005) also pointed out the problems that arise from a less participatory role for international planning practices in China. Friedmann (2005, p. 114) argued that ‘there are calls, chiefly from scholars mindful of international practice, for a more participatory approach to planning that goes beyond the interests of powerful stakeholders and their personal …relations with gatekeepers in the government’. Such a need for a more participatory approach in international planning practice, however, is challenged by the typical approach of using international planning practice as ‘concept plan[ning]’ (Wu et al. 2007, p. 180) and ‘design-based plan[ning]’ (2007, p. 182) in China. Friedmann (2005, p. 115) indicated that ‘urban governance in China is in a state of constant experimentation and flux’. These discussions suggest that while changes to China’s urban governance have provided opportunities for international planning practices, often these international planning practices have merely delivered concepts, rather than fully participated in implementation and project materialization. Challenges from Chinese ways of operation also affect the levels of international participation in planning and design.

In addition, international planning and design practices have been recognized more as by architects than landscape architects. While Wu et al. (2007, p. 183) list foreign architectural firms such as GMP and SOM as ‘renowned planning institutions’, this fact in itself reveals that existing discourses of international planning and design practice are largely centred first on architecture, then on urban design and new city planning. By contrast, planning and design work provided by international landscape architectural firms is discussed less. In relation to the argument of Wu et al., of place promotion in China for sustaining local growth, the recognition of projects designed by the world’s top architects in China suggest an expectation there that these international architectural practices will beautify the urban landscape. It also suggests a focus on architectural forms in inviting international planning and design services.

Western scholars' observation of China's urban development shows an attitude contrasting with the attitudes of the CHSLA that seek a shared platform between ancient/traditional Chinese methods and those of the modern West. Western observers appear more cautious about approaching the differences rooted in China's long history and the risks of deploying modern Western experience and values in contemporary China.

In his speech presented to Nanjing Forestry University in China, Eugene C. Hargrove (2005), an American professor and scholar in environmental philosophy, discussed the role of Western environmental ethics in non-Western countries. Hargrove (2005, p. 14) argued that the assimilation of Western environmental ethics in non-Western cultures needs 'tacit' and 'personal' knowledge, as ethics is based on the judgment of individuals rather than the judicial system. Hargrove (2005, p. 13) suggests that the delivery of Western environmental ethics outside the West needs to avoid aspects that remain controversial in Western society itself, such as the 'anthropocentric and non-anthropocentric distinction', 'the notion of intrinsic value', and 'the concept of natural rights'. Hargrove's argument shows tacit knowledge to be an important dimension in comparative studies between China and the West. It also suggests that understanding such tacit knowledge is fundamental to international landscape architectural practice, as it is increasingly involved in the transformation of valuable lands and resources affected by China's urbanisation. Furthermore, environmental ethics, as a Western concept, is increasingly introduced in the CHSLA.

Similarly, the validity of using western theory in China is questioned by Weller, the Australian landscape architect and educator. In his journey from the global city of Shanghai to the classical city of Suzhou, Weller observed the chaotic urbanizing fringes between the two cities. Weller (2005) questioned why the Eastern philosophy has never been revealed clearly in the products of China's rapid modernisation, despite its supposed sensitivity to natural and cultural systems. Weller argued (2005, p. 25) that landscape architectural practice in China has been a global production and that it is necessary to build cooperation between Chinese and Western landscape architects to ensure the achievement of professional standards that respond to national identity, ecology and art. Weller suggested that such cooperation might imply a major clash between Eastern and Western cultures. He sees the 'China project' as one of the biggest landscape architectural projects in human history, as the changes and landscape construction currently occurring in China are of a vast scale and have no precedents.

Compared with nostalgic picturesque landscapes manifest in the West, however, Weller (2005) observed that the urbanising fringe of China rarely shows a similar inclination to remember the lands lost to industrialisation and urbanisation or linked to the remote homeland of settlers, as is typical in Europe, America and Australia. Weller (2005, p. 25) commented that it is difficult for a foreigner to understand the changes occurring in China, despite the similarity of the urban outcomes resulting from 'constructive destruction' in Western terms. In relation to Western experience that might prove useful to China, Weller (2005, p. 26) pointed out that 'critical regionalism' in terms of the architecture and 'genius loci' of landscape architecture are the most important theoretical paradigms that have been imported to China, and yet they appear not to be successful. This further shows the perceived problem of transplanting western theories in China's reality of urban transition. Weller also suggested that large-scale planning is especially challenging in China, again highlighting the increasingly important role of planning in Chinese landscape architecture.

From the perspective of Chinese local design scholars, there have been similar concerns, while focusing more on western landscape forms popular in contemporary China. In the article, *Globalization and the Culture of Chinese Landscape*, Ye (2003, p. 6) identified two trends that he considered would be harmful to contemporary landscape architectural practice in China. One is using 'aesthetic names to recall the picturesque and poetic mode of traditional Chinese parks and gardens, but without real content'; the other is 'the copy[ing] of Western forms but without knowing the meaning underpinning those forms'. Ye's discussion suggests the unsuccessful outcome of using the Chinese garden tradition and Western forms in the designed and constructed contemporary landscape in China.

Contrasting with the critique of the omnipresent Western forms in the Chinese landscape, Ryckmans (1989) suggests the indifference of maintaining material forms as a rooted tradition in China. In his article, *the Chinese Attitude towards the Past—Failure of the Chinese to Maintain and Preserve the Material Expressions of their Culture*, he (Ryckmans 1989, pp. 2–3) pointed out:

> In the Chinese landscape, there is a material absence of the past – that can be most disconcerting for cultivated Western travelers – especially if they approach China with the criteria and standards that are naturally developed in a European environment. In Europe, in spite of countless wars and destruction, every age has left a considerable amount of monumental landmarks…In China, on the contrary, if we except a very small number of famous ensembles (the antiquity of which is quite relative), what strikes the educated visitors is the monumental absence of the past. Most Chinese cities – including, and especially those which were ancient capital cities and prestigious cultural centers…appear strangely devoid of traditional character. On the whole, they seem to be a product of late 19th century industralisation.

Ryckmans (1989, p. 3) concluded that 'the past which continues to animate Chinese life in so many striking, unexpected and subtle ways, seems to inhabit the people rather than bricks and stones. The Chinese past is both spiritually active and physically invisible'. Ryckmans (1989, p. 16) further pointed out that 'the modern Chinese intellectuals, progressives and revolutionaries have increasingly felt strangled by the seeming invincibility and pervasiveness of tradition'. The above discussion shows that the Chinese tradition lies largely in the ways of the life of Chinese people, rather than in material forms. This explains, in part, the difficulty of both international and local design practitioners to find solutions to continue Chinese traditions through the material forms of buildings, cities and landscapes and in the physical construction of the urban environment.

5.5 Conclusion

China's demand for international practice in landscape architecture is defined in a number of ways. Generally, such a demand was underpinned by the reform policies that were established in 1978. The consequent rapid economic growth and urban development, changes to society and urban governance, and the new market conditions for planning and design. International landscape architectural practice has always been a part of international architectural and planning practices in China. However, this was not clearly recognised until the late 1990s, when residential development was booming in China and international design firms came from the world over. International planning and design practice was initially invited to introduce Western experience, theories and models for urban development, and for the construction of "the new". It was rarely expected to preserve "the old and traditional". In such a context, the good intentions of some international practitioners in responding to sense of place remain unsuccessful.

Specifically, international landscape architectural practice in China has been promoted by the transformation of landscape architecture as a profession and discipline in China, especially between 1995 and 2006. It was during this period that vast rural lands were being transformed into urbanised use; while expertise in landscape planning and environmental strategies remains insufficiently developed in local profession.

The initial recognition of international residential landscape design suggests the demand in China for new landscape forms in relation to socio-cultural changes, akin to culturally sensitive practice. The recent engagement in waterfront projects also shows the demand for environmentally sensitive practice. The above two hints of international landscape architectural practice in China, however, contrast with the debate of the qualities of sense of place of the many designed and constructed landscapes in China. The spiritually active and physically nebulous Chinese traditions suggest that ways to sustain the characteristics of place in the development of Chinese cities depends greatly on the understanding of their people. This situation is unique to China and may differ from the experience of international practitioners from the West.

References

Benewick, R., & Wingrove, P. (Eds.). (1999). *China in the 1990s*. Hampshire, Macmillan: Basingstoke.

Chen, B., & Ge, Z.-Q. (2006). YU Kong-jian: Landscape architecture of China is inevitably of the world. China Green Times. Retrieved February 06, 2009, from http://www.landscape.cn/News/event/2006/20061030795.html.

Foot, R. (1999). China's foreign policy in the post-1989 era. In R. Benewick & P. Wingrove (Eds.), *China in the 1990s* (pp. 234–244). Basingstoke, Hampshire: Macmillan.

Friedmann, J. (2005). *China's urban transition*. Minneapolis: University of Minnesota Press.

Gerkan, M. V. (2008). *Projects in China: architects Von Gerkan*. Munich, Prestel: Marg and Partners.

Hall, D. L., & Ames, R. T. (1995). *Anticipating China: Thinking through the narratives of Chinese and Western culture.* Albany: State University of New York Press.

Han, B.-Y., & Shen, S.-X. (2005). Landscape architecture design based on regional feature. *Chinese Landscape Architecture* (7), 61–67.

Hargrove, E. C. (2005). The role of western environmental ethics in non-western countries: West Meets East. *Journal of Nanjing Forestry University (Humanities and Social Sciences Edition), 5,* 11–16.

Lau, P. (2004). EDAW: The integrated world. *Chinese Landscape Architecture* (1), 12–21.

Li, B. (2005). Projects designed by foreign architects in Beijing since 1978. *Time+Architecture* (1), 14–18.

Lin, J. (1994). *The opening of the Chinese mind: Democratic changes in China since 1978.* Westport, Conn., Praeger

Lin, X. (2003). The belt collins phenomenon. *Chinese Landscape Architecture* (3), 47–49. (林潇. 贝尔高林现象).

Liu, K. (2004). *Globalization and cultural trends in China.* Honolulu: University of Hawai'i Press.

Liu, J.-Q., Wang, B.-L., et al. (2004). Opinion of 'landscape architecture in terms of the wholeness of land'. *Chinese Landscape Architecture, 07,* 41–44.

Liu, T.-F. (2000). A child lack of critique-modern chinese landscape architecture. *Chinese Landscape Architecture, 16*(5), 26–28.

Lou, Y.-Y. (2005). Chinese Architecture: Intergrowth of the inside and the outside. Retrieved December 06, 2007, from http://www.landscape.cn/news/html/news/detail.asp?id=26434. (12 Aug 2005).

Ryckmans, P. (1989). The Chinese attitude towards the past–failure of the Chinese to maintain and preserve the material expressions of their culture. Papers on Far Eastern History, The Australian National University (pp. 1–16).

Ruano, M. (1999). Strangers in…Paradise? Instant China: notes on an urban transformation. Barcelona, Editorial Gustavo Gili: 30–88.

Mackerras, C., Taneja, P., et al. (1994). *China since 1978: Reform, modernisation and 'socialism with Chinese characteristics'.* Melbourne, New York: Longman Cheshire; St. Martin's Press.

Meng, Z.-Z., & Wang, X.-R. (2002). Recovering the discipline of landscape planning and design. Retrieved January 15, 2009, from http://www.landscapecn.com/paper/detail.asp?id=28.

Wang, F. (2007). Application of New Regionalism in Urban Landscape Design. *Chinese Landscape Architecture, 23*(12), 60–63.

Wang, Y.-C., & Liu, J.-G. (2007). The Dialogue between modern landscape and regional historical culture-The historical culture protection design for Xujiahui Park. *Chinese Landscape Architecture* (7), 43–46.

Weller, R. (2005). On the way to Suzhou-observations of contemporary landscape architecture in China. *Chinese Landscape Architecture, 21*(116), 24–26.

Wei, D.-Y. (2006). A Chinese-english dictionary (Revised Edition). English Department of Beijing Foreign Language University. Beijing, Foreign Language Teaching and Research Press.

Williams Asselin Ackaoui & Associates. (2002). Shanghai Allegory Park. Retrieved Febrauary 09, 2009, from http://www.waa-ap.com/an/projetsan/internat/xuji/XuJiaHuian.html.

Wu, F., Xu, J., et al. (2007). *Urban development in post-reform China: state, market, and space.* London; New York: Routledge.

Wu, Y.-P. (2004). Urging the establishment of the registration system of landscape architects: A joint letter for an unsolved problem of nine years. Retrieved April 09, 2008.

Xue, C. Q. L. (2005). *Building a revolution: Chinese architecture since 1980.* Hong Kong, London: Hong Kong University Press; Eurospan [distributor].

Yan, G.-T., & Lu, Y. (2008). A reflection over China's modern landscape diversity under the context of globalization. *Chinese Landscape Architecture* (10), 14–18.

Ye, J. (2003). Globalization and the culture of Chinese landscape. Guangdong Landscape.

Yu, K. J. (2000). Searching site property: a few channels of landscape design and their comparative study. *Architecture Journal, 3*(2), 45–48.

Yu, K.-J. (2004). Landscape architecture in terms of the wholeness of land. *Chinese Landscape Architecture* (07), 37–40.
Yu, K.-J. (2006). Walking towards new landscapes. *Architectural Journal* (5), 73.
Zhang, L. (2006). Rethinking of the identity crisis in Chinese traditional gardens. *Chinese Landscape Architecture* (9), 60–63.

Chapter 6
The Li Lake Area and the Proposed Development

Abstract This chapter explains the regional context and history of the Li Lake area, and the proposed development that involved international landscape architectural services and Australian landscape architects. This chapter identifies the regional and local character of the Li Lake area from literature, and the scope of the development and clients' motivations to invite international planning and design services. This chapter relates to Chap. 7 that studies the process of providing services and to Chapter Eight that studies the interpretation of the characteristics of the Li Lake area in various planning and design outcomes. This chapter has three major sections. Section 6.1 discusses the regional context of the Li Lake area, including the Jiang-nan region and the Lake Tai plain. Section 6.2 examines the ancient and recent histories of the city of Wuxi and Li Lake. Section 6.3 analyses the scope and structure of the case project. It also outlines the expectations of the international design teams and the phases of international design practice.

Keywords Li lake area · Local characteristics · Local government · International design teams · Project process

6.1 The Regional Context: Jiang-Nan

The Li Lake area is situated in a unique region of China: Jiang-nan, where the natural environment has nurtured a prosperous economy for at least a thousand years. During this long period of prosperity, the intellectual life of Jiang-nan was recognised as outstanding throughout China and deeply influenced Chinese culture as a whole. The prominent economic and cultural position of Jiang-nan in Chinese history was underpinned by the benign climate and abundant natural resources of the Lake Tai plain, which is the core area of Jiang-nan (Feng 2007: 64, 67). Lake Tai is the third-largest freshwater body in China and the largest in the Jiang-nan region. The economic prosperity has led to the most rapid urban development of Jiang-nan and Lake Tai plain in recent decades.

Regarding the term Jiang-nan, 'Jiang' means Yangtze River and 'nan' means the south. Geographically, Jiang-nan refers to a region in the lower Yangtze valley,

Y. Zhang, *Place Making in International Practice of Landscape Architecture*, Advanced Topics in Science and Technology in China 64,
https://doi.org/10.1007/978-981-16-2442-1_6

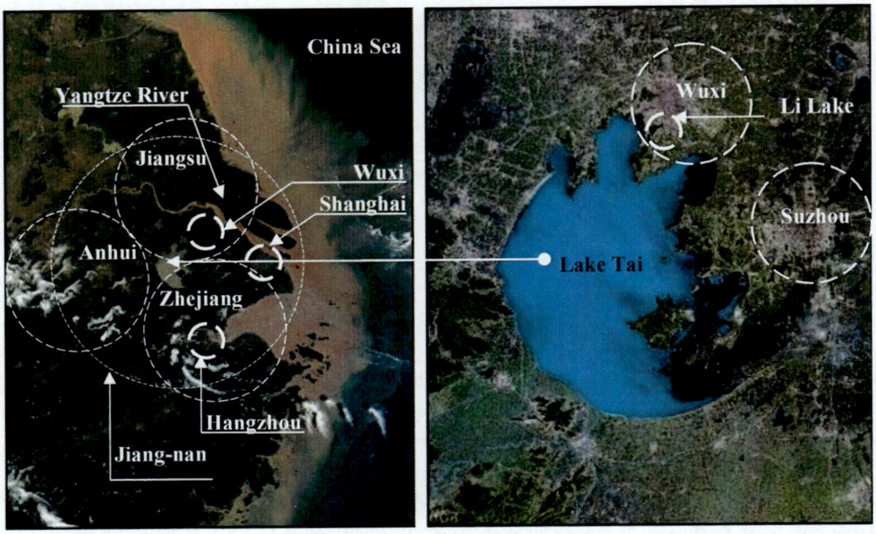

Fig. 6.1 Location of Jiang-nan and its key cities. *Sources* Left—http://a2.att.hudong.com/74/96/01300000202579122336969185580.jpg; right—http://baike.baidu.com/view/1596.htm; retrieved 30 May 2010, notation by author

including southern Jiang-su and An-hui and northern Zhe-jiang provinces (Fig. 6.1, left), that is much celebrated in poetry for its beauty and the joys it provides (Wei 2006). Figure 6.1 shows Wuxi, the municipal city of the Li Lake area, neighboured by Lake Tai ten kilometres away. The city of Suzhou, renowned for its traditional private gardens, is forty-five kilometres away. Another city of gardens, Hangzhou, is about 150 km distant. The global city of Shanghai, the major port for foreign tourists in Eastern China, is 128 km from Wuxi (Fig. 6.1). The development of the Li Lake area is closely associated with improving Wuxi's local tourist economy by attracting tourists from not only Jiang-nan but also other parts of China and the rest of the world (RDUDP 2002; EDAW 2002 August; Sun et al. 2007; Sun and Shao 2007; WMCPC 2007).

Jiang-nan is not the name of an administrative area or a geographical area with fixed boundaries. Rather, it is a culturally defined concept. From the perspective of historians, the extent of Jiang-nan is unfixed (Fig. 6.2). Since the Qing dynasty (221–206 B.C.), Jiang-nan's expanse has shifted from west to east and shrunk substantially. During the ancient period from the Tang (618–907) to the Ming (1368–1644) dynasties, the area conceived of as being Jiang-nan came to centre on the Lake Tai plain (Zhou 2007). Jiang-nan's prominent economic and cultural position in China began during the Southern Song (1127–1279) dynasty, when the emperor Zhao Gou—a painter and poet instead of a politician—fled his Northern capital from the invasion of northern nomadic tribes and established a temporary capital in Hangzhou. Following the emperor, intellectuals of high achievements concentrated in Jiang-nan.

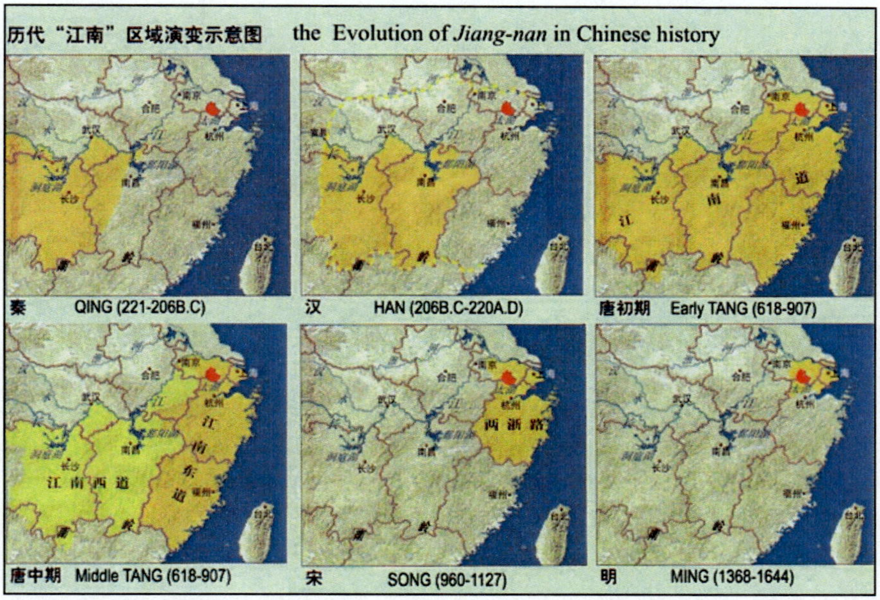

Fig. 6.2 The central location of the Lake Tai plain in Jiang-nan. *Source Chinese National Geography: Jiang-nan special* (Zhou 2007: 61), notation by author

Jiang-nan's prominence increased during the Ming dynasty, when the emperor Zhu Yuan-zhang established a capital in Nanjing about 150 km from Wuxi.

In Fig. 6.2, the yellow area denotes the scope of the Jiang-nan region throughout Chinese history; the red area indicates Lake Tai. The figure shows that Jiang-nan has changed over time and the central position of the Lake Tai plain was established in the Ming period.

The regional sense of place of Jiang-nan is expressed through the forms of its villages and towns, buildings and gardens. Throughout its history, Chinese literature has identified cities such as Suzhou, Hangzhou, Nanjing, Wuxi, and Shanghai with Jiang-nan. A prominent aspect of the literary description of these cities is their water villages and towns that often spread along rivers. In these villages and towns, shops, teahouses and residences accumulate along the riverbank, and all sorts of entertainment occurs along traffic nodes, such as bridges and jetties. While Chinese literature identifies the above cities with Jiang-nan, these cities are also renowned for their gardens, which signify the literary culture of Chinese intellectuals. Figure 6.3 shows that rivers and lakes have greatly influenced spatial forms in Jiang-nan.

The regional sense of Jiang-nan is also expressed through its people and their customs and practices of everyday life. The people of Jiang-nan are distinguished from both northern and southern Chinese people. This is a result of its mild natural conditions and prominent literary culture. In his book, *My Country and My People*, Lin (1936) compared Chinese people in different regions. He (1936: 18) described

Fig. 6.3 Jiang-nan's regional sense of place—villages, towns, gardens. *Sources Chinese National Geography: Jiang-nan special* (2007: 63, 118, 131)

the northern Chinese people as 'acclimatised to simple thinking and hard living, tall and stalwart, hale, hearty and humorous', while people of Jiang-nan as:

> …inured to ease and culture and sophistication, mentally developed but physically retrograde, loving their poetry and their comforts, sleek undergrown men and slim neurasthenic women, fed on birds'-nest soup and lotus seeds, shrewd in business, gifted in *belles-letters*… [They are] off[-]springs of the cultured Chinese families who crossed the Yangtse [Yangtze] with their books and paintings during the end of the Chin Dynasty [265–410], when China was over-run by barbaric invaders.

The more sophisticated and feminine character is rooted in the people of Jiang-nan, and has been for thousands of years. Although such character may be less obvious in modern decades, the prominent economic and cultural position of Jiang-nan cities in contemporary China is the evidence of the unique contribution of the people of Jiang-nan. Such a character is also reflected in their smart ways of using rivers and lakes to achieve prosperity and have recreation. The following figures show how the people of Jiang-nan conduct business and make a living (Fig. 6.4), and reside and entertain along rivers and lakes (Fig. 6.5).

Fig. 6.4 The characteristics of Jiang-nan—everyday working. *Sources Chinese National Geography: Jiang-nan special* (2007: 44, 141)

Fig. 6.5 The characteristics of Jiang-nan—everyday life. *Sources Chinese National Geography: Jiang-nan special* (2007: 83, 93)

As a culturally prominent area, its traditional gardens and parks have been emphasized as the symbol of Jiang-nan. These gardens and parks were outcomes of cultural practices led by the upper classes comprising high officials, literary men and established merchants in the past time. The benign climate, advanced economy and the easy transportation offered by rivers and canals also promoted such cultural practices in Jiang-nan and consequently brought renown to the cities of Hangzhou, Suzhou and Yangzhou on account of their gardens and parks (Peng 1988: 50–52; Zhou 1990: 253; Yang 1994: 10–11; Fang and Qian 1999: 10–11; Pan 2001: 2–3). The Chinese term *Jiang-nan yuan-lin* indicates how Jiang-nan is popularly associated with these traditional landscapes.

Of the prominent Jiang-nan gardens, most were walled private gardens owned by the upper classes. Physically, they are characterised by architectural features such as houses, moon doors, pavilions, bridges and rockeries (Figs. 6.6 and 6.7). Experientially, the gardens are characterised by moral, poetic, and picturesque sentiments, expressed through tablets engraved with verses, through calligraphic couplets (Keswick 1986: 147–153), and through the composition of gardens that imitate those depicted by Chinese landscape paintings (Fig. 6.8). The rich experience of the gardens also arises from the layers of garden spaces when the intricate components of the gardens unfold along with the footpath.

The above images of Jiang-nan gardens present an omnipresent feature: the Chinese rockery and the use of Lake Tai rock. Keswick (1986: 155) described how western visitors would confront these rocks: 'Grotesque and grey, tortuous and massive, they dominate the scene in much the same way that flower beds highlight an English garden'. The worship of rocks in these gardens, as explained by Keswick

Fig. 6.6 Jiang-nan traditional gardens—dominant architectural forms. *Source Chinese National Geography: Jiang-nan special* (2007: 120, 123, 125). Left—*Zuo-zheng* (means humble politician) Garden of Suzhou, of which the design followed the typology of the site. Middle—*Xiao Fei-hong* (means little rainbow) Bridge of *Zuo-zheng* Garden, viewed from a garden house. Right—*Yi* (means art) Garden, the smallest garden of Suzhou, which was first established in 1541 and is less than 4000 square metres

Fig. 6.7 Jiang-nan traditional gardens—dominant rockeries. *Source The Chinese Garden* (Keswick 1986: 117)—a drawing of Chia family garden in Nan-jing from the 18th Century novel *Dream of the Red Chamber*

(1986: 155–161), implies the worship of the spiritual realm of Taoism, as the physical character of Lake Tai rocks symbolise such a spiritual realm.

The traditional parks[1] of Jiang-nan are typically along the outskirts of towns at a distance from everyday urban lives. These parks are also found among great waters and mountains. The Li Lake area falls in the categories of both outskirts and great waters. In his studies of the art of landscaping of Jiang-nan, Pan (2001: 249–282) indicated that the cultural landscape of the outskirts of towns was formed by three forces. One force was large land transformations for the purposes of flood control, quarrying and dredging, which were often led by the official literati.[2] The second force was the great water and mountain areas, such as Lake Tai, that are often associated with the immortal world and religious beliefs, such as Buddhism and Daoism, and identified with the spiritual realm of wise men. The third force was the legends of heroes and beauties, and also the paintings and poems of famous intellectuals that related great deeds or events to particular landscapes and moral teaching. These three

[1]As discussed in the Introduction Chapter, the Western term 'park' and its modern recreational meanings were introduced to China in the late nineteenth and early twentieth centuries, along with the Western colonisation and the building of parks in leased territories such as in Shanghai and Hong Kong. In this section, the term 'park', however, refers to the public domain of the landscaped area in traditional Jiang-nan towns and villages, which developed in their own right for purposes related to flood control, the quarry industry, or religion, and everyday outdoor use. The difference from the meaning of park in Western terms is further discussed in Chap. 9.

[2]In ancient China, intellectuals were required to pass several exams to become government officials. These exams centred on literary subjects. Therefore, in Chinese history, officials were often skilled in calligraphy, painting and poetry, and often renowned for their literary achievements.

Fig. 6.8 Jiang-nan traditional gardens—the literary experience. *Source The Chinese Garden* (Keswick 1986: 108, 146, 149, 153)

forces suggest an upper-class influence during ancient times in shaping the traditional park of Jiang-nan. Figure 6.9 shows that similar to traditional private gardens, the cultural expression of traditional parks exists not only in architectural features such as moon doors, towers, pavilions and footpaths, but also in texts engraved in rocks that tell stories. The top left image shows a park in the outskirt of Suzhou, in which the calligraphy was made more than 1500 years ago. The calligraphy shows the name of the pond—Sword Pond (means water for washing your sword). The top right image

Fig. 6.9 Jiang-nan traditional parks—the outskirts. *Source The art of landscaping in Jiang-nan* (Pan 2001: 255, 272, 279)

shows a landscape shaped by quarry and lakeside temple in the town of Shaoxin. The bottom image shows classical garden elements presented in the banks and islands in West Lake of Hangzhou. These dams were originally made of silt dredged from the lake in the Ming dynasty.

The traditional gardens and parks of Jiang-nan have been significant components of the cultural landscape there, symbolising the literary culture that dominated Chinese history and in turn influenced Chinese culture as a whole. Most of these gardens, once privately owned, are now open to the public and are places of interest to tourists. The iconic features of the gardens, parks, towns and villages of Jiang-nan are also referenced in China's contemporary landscape architectural practices. Figure 6.10 shows how these elements, especially architectural forms, are expressed in recent residential landscapes (Fig. 6.10, top) and university campuses (Fig. 6.10,

Fig. 6.10 Influences of the traditional forms on contemporary design. *Sources* Top—http://www.coolsc.net/3D-Details/2007/08/24/18891-13.htm#2, retrieved 30 April 2010. Bottom—photographed by author

bottom) in China, such as moon doors, rockeries (Fig. 6.10, bottom left) and arch bridges (Fig. 6.10, bottom middle).

6.2 The History of Wuxi and Li Lake

The Li Lake area refers not only to the waterfront of Li Lake but also to the area between the northern shore of Li Lake and the existing city of Wuxi. As a southern city of Jiang-su Province, Wuxi is bordered by the Yangtze River on the north and Lake Tai on the south. It is the largest city along the northern shore of Lake Tai and the city closest to the lake (Zhuang 1995). Wuxi has a total area of 4787 km^2 and features 'an intra-town hill, a winding lakeshore and a canal running through' (WMCPC 2007: 5). The dark area in Fig. 6.11 refers to hills, which surround the flat land of the Li Lake area and Li Lake. In 2002, Wuxi had a total population of 3.8 million, with 1.3 million people residing in the urban area (RDUDP 2002). As a part of Lake Tai that intrudes into the municipality of Wuxi, Li Lake is separated from the broader body of water mainly by rolling hills to the southwest (Zhuang 1995: 264). Li Lake consequently has the characteristics of an isolated enclosed lake, a fact that has caused severe water problems.

The following images show that along with the waterfront of Li Lake, there are hills and historic landscapes, such as Turtle Head Islet, categorised by the Chinese government as a four-star scenic area, Li garden (built in 1920) and Ji-chang Garden

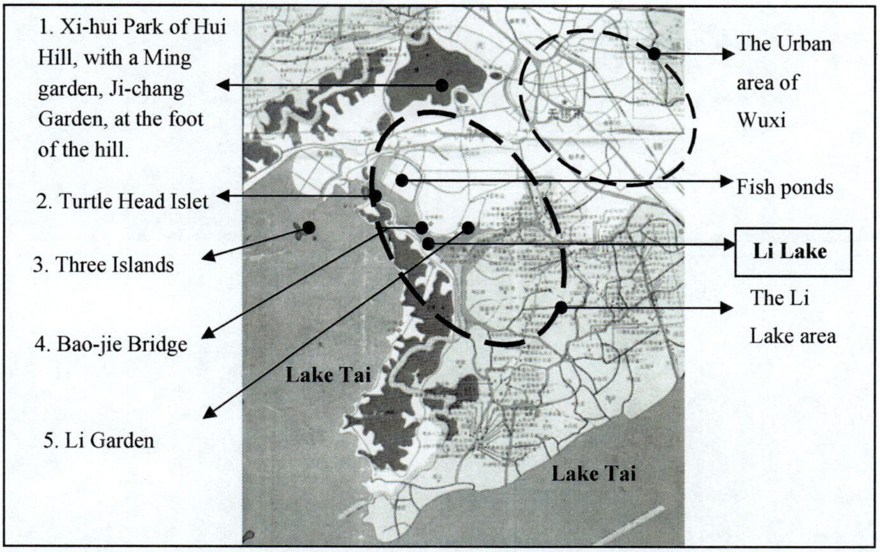

Fig. 6.11 Location of the Li Lake area and its historic landscapes. *Source* Wuxi Map (Jiang-su Survey Bureau 1996), notation by author

(built about 500 years ago). On the other side of Turtle Head Islet lies Three Islands in the grand water body of Lake Tai. The name of Three Islands carries a Taoist meaning: the land of eternal life. Such a feature is typical in traditional Chinese gardens and parks. These historic landscapes had attracted seven million visitors each year (Fig. 6.12).

The civilisation of Wuxi has a history reaching back more than 2500 years. Figure 6.13 shows how the structure of the urban area had formed by 1268. The form remains approximately the same to the present day (Fig. 6.14). The canal system

Fig. 6.12 Historic landscapes around the waterfront of Li Lake. *Source Wuxi Local History* (Zhuang 1995: front pages), notation by author

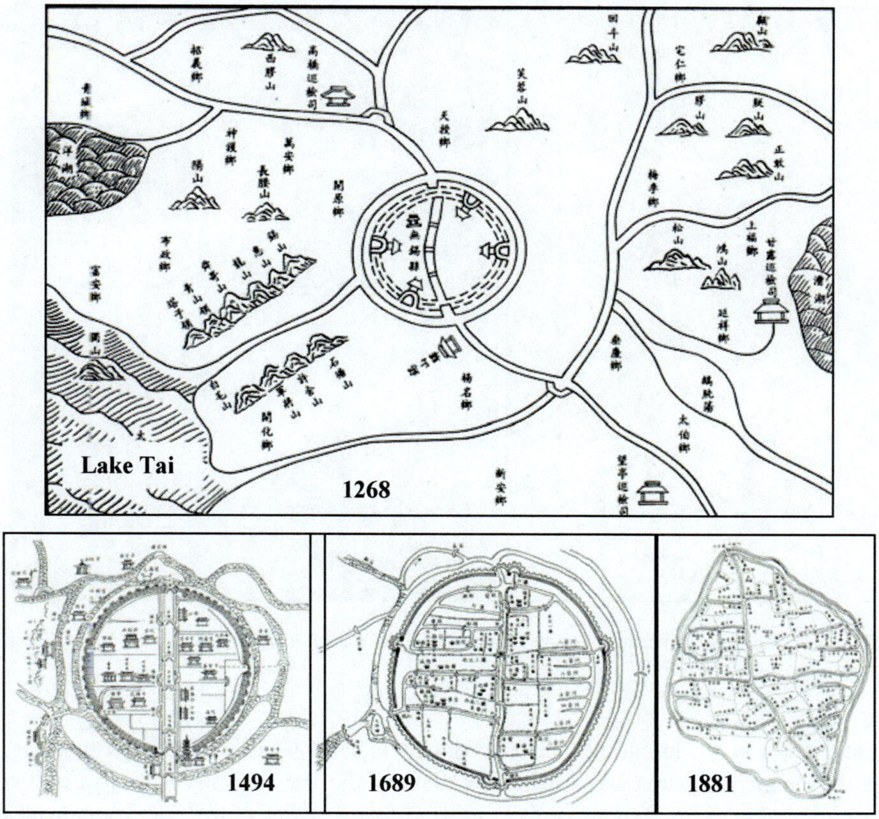

Fig. 6.13 The surviving ancient urban structure of Wuxi. *Source Wuxi Local History* (Zhuang 1995: 126, 128, 129, 130)

greatly influenced the survival of this urban form. Figure 6.13 shows the increase in the number of canals since 1494 that surrounded or connected to the central canal shown in the map of 1268. These canals include the renowned Grand Canal, which was built 1400 years ago to connect Beijing and Hangzhou and has a length of nearly 1794 km. This canal is still in use. The pictorial record from 1881, however, shows a change from the original circular pattern to a form closer to that depicted in present-day maps, as shown in Fig. 6.14. This suggests that in ancient times such pictorial records were based on experience and memory, but since 1881 modern technology of survey started to be used to record the city. This modern type of map became a reference for present planning and design practice.

The local characteristics of Wuxi and nearby areas are not only embodied in physical gardens, parks and urban forms, but also have been expressed in the region's underlying folk culture and Wu culture (Sun et al. 2007: 32). The folk culture refers to Wuxi as an area full of legends and folklore. For instance, of the name of Wuxi, 'Wu' means none; 'xi' means tin. Wuxi literally means no tin. A popular local legend shows

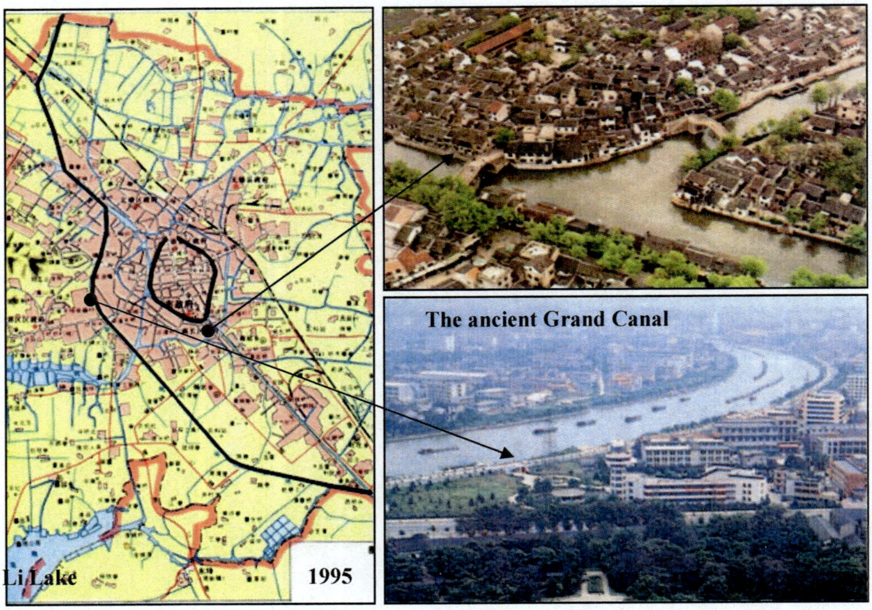

Fig. 6.14 The existing urban structure of Wuxi. *Source Wuxi Local History* (Zhuang 1995: front pages), notation by author

that the area was originally called You-xi, by which 'You' means having something. When the tin in the area was dug out, the name of the area became Wuxi. In addition to stories of Wuxi being a place full of tin, there are stories of historical personages such as Zhang-bo, a hero of flood control, and Fan Li, a politician who retreated to Li Lake, wrote the first book in Chinese history on fish farming, and lived with Xi Shi who was one of the four famous beauties in Chinese history.

Wu culture is characterised by its influence over the mercantile culture of Wuxi (WMCPC 2007: 47; Yi 2008), as Wuxi was a seedbed of national industry in China (Zhuang 1995: 4). The established local merchants and industrialists of the early twentieth century and their contributions to the modern development of Wuxi composed a significant part of the local history and characterised its mercantile culture. These stories are also about outstanding literary men and women, painters and poets who practiced in the area. Their literary and artistic works formed a significant part of the local culture. Figure 6.15 shows two ancient paintings of the lake area. The top images are two parts of a vertical hanging scroll painting. The painting illustrates Hui Park of the Li Lake area and was made in the seventeenth century. The bottom image is a painting of Lake Tai made in the fifteenth century, which shows the Three Islands of Lake Tai. Both paintings contain an arch bridge, of which the style has become a reference to local history in the newly constructed Western Li Lake waterfront.

Fig. 6.15 Ancient paintings of the landscape of Li Lake and Lake Tai. *Source* Top—*Wuxi Local History* (Zhuang 1995: front pages). Bottom—*Wuxi stepping toward an ecological water-mountain waterfront city* (Sun and Shao 2007: 10)

The abundant natural resources of the Lake Tai plain and the water control conducted by prior inhabitants made Wuxi an area rich in fish and rice production (Zhuang 1995: 2). Because of the easy transport enabled by its canals, Wuxi has been historically acknowledged as a trading centre for rice and cloth, particularly from the sixteenth to the eighteenth centuries (1995: 2). In 1911, Wuxi already had twelve factories, including cotton mills, flour mills, rice-hulling mills, silk-reeling mills and dyeing mills (1995: 4). Twenty-five years later, the number of factories had increased to 315, resulting in the area being called ‘the mini Shanghai’ (1995: 4).

The region's industrial history has had an extraordinary impact on the environment. It has reclaimed land from the lake waters for rice paddies, dredged canals for irrigation and for exporting local products, directly discharged pollutants into rivers and canals, and, in the late twentieth century, developed fish farming. These activities have not just worsened the water quality of Lake Tai and Li Lake; they have reduced the water surface area significantly. Before 1949, Lake Tai had a surface area of 2500 km^2; it is now 2338 km^2 (Zhang 2007: 136). Similarly, the low-lying waterfronts of Li Lake were gradually transformed into croplands and fish ponds (Fig. 6.16). Prior to 1995, Li Lake had an area of 9.5 km^2. By 1995 this had shrunk to 8.6 km^2 (Zhuang 1995: 264), and by 2002 it was 6.4 km^2 (WMCPC 2007: 13).

Fig. 6.16 The recent history of the Li Lake area. *Sources Unveiling the beauty of Li Lake: Wuxi well on its way towards ecological well being* (WMCPC 2007: 23, 29,31); *Wuxi stepping toward an ecological water-mountain waterfront city* (Sun and Shao 2007: 14, 21), notation by author

Fig. 6.17 Traditional local activities—swimming and fishing. *Source Wuxi Local History* (Zhuang 1995: front pages)

In recent decades, the growth of Wuxi's tourism economy has also resulted in uncontrolled land development, typified by the building of a European-style park for film shooting. During the late 1990s, the park, located in the central area of the northern shore of Li Lake (Fig. 6.16: 2), suffered a loss of interest among tourists. The following figure also shows fish ponds, pig farms and houses of peasants along the waterfront of Li Lake (Fig. 6.16: 1, 3, 4 and 5). These all became sources of pollution in Li Lake and Lake Tai.

Despite water pollution in both Li Lake and Lake Tai, traditional local activities such as swimming and fishing remained popular (Fig. 6.17). However, the explosion of blue algae in Lake Tai in the 1990 s not only affected these activities but also brought about a crisis in the provision of drinking water for Wuxi residents. Wuxi, the biggest city to the north of Lake Tai, has been identified as one of the two major sources of one-third of the lake's pollution (the Ministry of Environmental Protection of China 2001: 4) (MEPC). Li Lake, located between Lake Tai and urban Wuxi, became the most polluted because of the direct discharge of agricultural, industrial and domestic pollution. Furthermore, Li Lake's partial enclosure by the southwestern hills and the construction since 1950 of dams and weirs to improve irrigation and control floods (Kahn 2007) have all slowed the circulation process that cleanses the lake, a process that now takes more than one year to achieve (MEPC 2001: 4).

On 25 October 2001, MEPC (2001: 2) issued 'the Tenth Five-year Plan of Lake Tai Pollution Prevention', in which Li Lake was targeted as the highest priority of the Five-year Plan from 2001 to 2005. During this time, two international design teams provided planning and design services for the development of the Li Lake area, and Australian landscape architects played key roles in these design teams. The following section describes the context and structure of the project. It also briefly discusses the reason for inviting the two international design teams, which are acknowledged by the government clients as foreign landscape architectural firms.

6.3 The Proposed Development

The clients (the Wuxi Municipal Government) titled the proposed project 'Planning and Construction of the Li Lake Area'. This project was a combination of environmental engineering to improve Li Lake's water quality and the previously proposed urban development of Wuxi to extend the urban area to the Li Lake waterfront. The environmental engineering was embedded in the larger environmental enhancement project to improve the water quality of Lake Tai. This project was funded by the central government of China (Sun et al. 2007: 7) and technologically supported by the Chinese Research Academy of Environmental Sciences (2007: 20–22). In terms of urban development, in 2000 the Wuxi Municipal Government set a target for adjusting Wuxi's administrative area (Sun and Shao 2007: 7). The aim was to expand Wuxi's urban area from 517 to 1662 km^2 (2007: 6)—that is, to more than triple it. Figure 6.18 shows the combined objectives of the project.

The environmental engineering components for the project were embodied in six sub-projects. A feasibility research report for the Comprehensive Water Treatment of Li Lake was prepared. This was provided by the Committee of Jiang-su Province Development and Planning and approved by the Committee of National Development Planning on 23 December 2002 (Sun et al. 2007: 7). These six sub-projects included (2007: 7):

1. Desilting 5.6 km^2 of water surface and 240 km^3 of silt in Li Lake
2. Replacing 2.5 km^2 of fish ponds
3. Replacing the water
4. Setting controls to the input rivers and canals
5. Renovating 19.4 km of Li Lake waterfronts
6. Rehabilitating the ecology of Li Lake, including planting 2.33 km^2 of water plants

The total cost of the six sub-projects was estimated at 540 million RMB (now equivalent to AU$90 million), 12% of which would be supported by the Chinese central government and nearly half of which was to be sourced from the Wuxi Municipal Government (Sun et al. 2007: 7). Except for the fifth sub-project (the renovation of the Li Lake waterfront), the sub-projects were related to the direct control of water quality and financially supported by additional funding from the

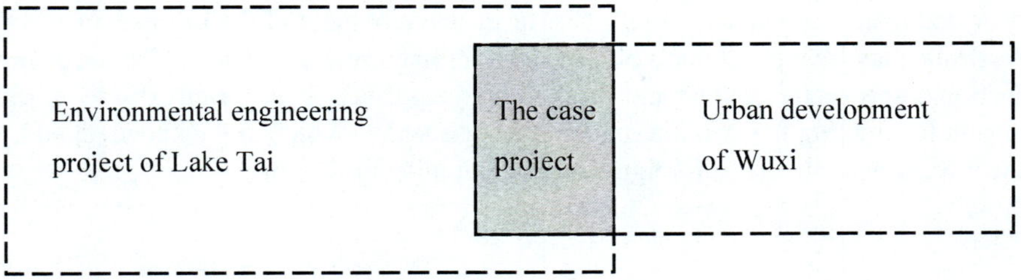

Fig. 6.18 The two components of the case project. *Source* Adapted by author from case data

Chinese central government (2007: 7). The Wuxi Municipal government aimed to restore the water surface of Li Lake from 6.4 to 8.6 km^2 (2007: 17).

The urban development component of the project was associated with the rapid growth of Wuxi's economy and urban population in the previous decade (Wuxi Statistic Bureau 2007), a situation similar to many other cities located in China's 'eastern coastal urban development zone' (Sun and Shao 2007: 1). Influenced by urban expansion, in 2001, the waterfront of Li Lake had already been less than ten to fifteen kilometres from the urban area of Wuxi. Therefore, the area between the northern shore of Li Lake and the existing city became the major focus of Wuxi's urban construction. The planned expansion of the urban area towards the lakeside and water bodies enabled the Wuxi Municipal Government to aim at creating what they described as a 'water-mountain' city and a 'lakeside' city (Sun and Shao 2007: 6). Following this, a principle for Wuxi's development was written into the Executive Proposal of the Wuxi Municipal Government for the period between 2002 and 2004 (2007: 6), which can be translated as: 'Branding Lake Tai, Eulogizing Grand Canal, Constructing a water-mountain city' (2007: 6). According to this principle, Lake Tai, the rolling hills surrounding Li Lake and the Grand Canal all became key aspects of Wuxi's future identity and were emphasised by the Wuxi Municipal Government. This principle, which guided Wuxi's urban development as a whole, also underpinned the development of the Li Lake area. Furthermore, another clear aim of the government was to transform the area into a well-known destination for both domestic and foreign tourists (RDUDP 2002: 4; EDAW 2002 August: 4; Sun et al. 2007: 8).

The Li Lake area identified for redevelopment was composed of the water body of Li Lake, the surrounding hills and the flat lands (Fig. 6.19). The area covers approximately 20 km^2; of this area, 48% is water body, 8% is hill area and 44% is flat land (Sun and Shao 2007: 7). The existing Bao-jie Bridge divides Li Lake into Western and Eastern Li Lake. The waterfront is a line approximately 36 km long and between 80 and 250 metres wide (2007: 7). The area between Li Lake's northern shore and the existing city was proposed for Li Lake New City, a cultural and economic centre connecting the old urban area with the lake. This central area was to cover six square kilometres. Figure 6.19 shows the scope of the Li Lake area, the location of the northern shore and the proposed new city.

The Wuxi Municipal Government further proposed five principles for the development of the Li Lake area (Sun et al. 2007: 16–19). The first can be translated as 'rigorously enforcing planning controls, opening the planning and design markets, raising the threshold for granting permission for construction' (2007: 16). The second was 'prioritising water control and tree planting for a friendly environment' (2007: 17). The third was 'prioritising road construction for the structure of a water-mountain city' (2007: 17). The fourth was 'building a new harmonious city, cherishing the peasants who lost their homes, lands and jobs for the development' (2007: 18). The last was 'integrating culture into the new city, constructing culture simultaneously with the construction of an ecological environment' (Sun et al. 2007: 18–19).

The implementation of the first principle was embodied in the decision of the Wuxi Municipal Government to use foreign planning and design services. Consequently,

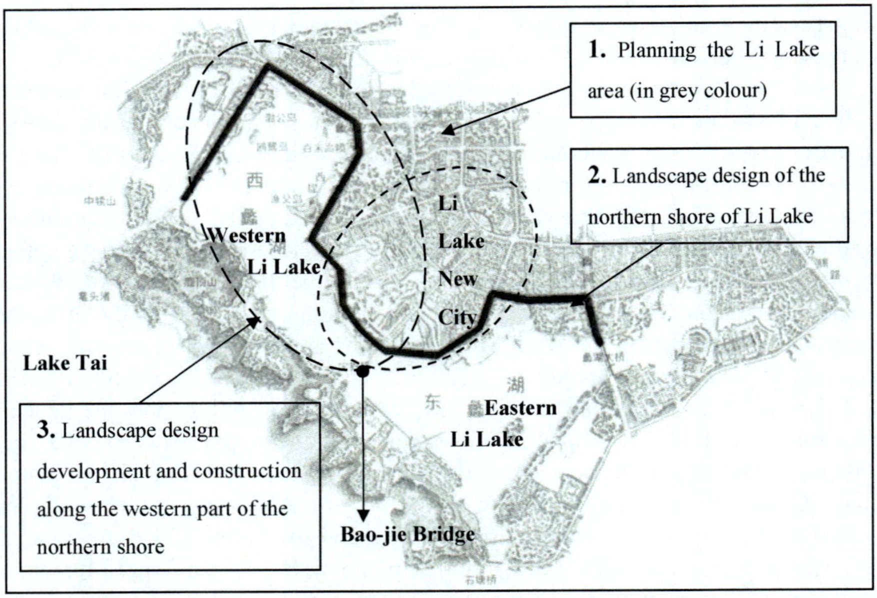

Fig. 6.19 Three phases of the case project. *Source Wuxi stepping toward an ecological water-mountain waterfront city* (Sun and Shao 2007: 20), notation by author

towards the end of 2001, two international design teams were invited to participate in the project. From the end of 2001 to the end of 2005, major work provided by the two international design teams covered three phases: first, planning of the 20 km^2 of the Li Lake area (Fig. 6.19: 1), with a focus on the new city area; second, landscape design of the northern shore waterfront that included both Western and Eastern Li Lake (Fig. 6.19: 2); and third, landscape design development and construction supervision of the northern shore waterfront of Western Li Lake (Fig. 20: 3). Both teams were involved in the planning phase, but only one team participated in the later phases. By the end of 2005, the construction of the Western Li Lake waterfront was completed. The area was also significantly transformed. Figure 20 shows two aerial photos depicting such changes, taken in 2001 and 2007, respectively. Bigger sized images are in Appendices.

In mid-2005, the Eastern Li Lake waterfront became the focus of the work of design and construction. After that, landscape design development has been dominated by professionals from the local area, such as the Wuxi Landscape Architectural Institute, and from nearby cities, such as a design institute from Zhejiang University (Guo and Li 2007). Foreign consultancy teams from other countries have been also involved. In mid-2005, the Canadian FK consultancy participated in the transformation of the Chang-guang Stream Wetland into an ecological park. The development of the Eastern Li Lake waterfront was funded by both the Wuxi Municipal Government and private investors. In contrast, the Western Li Lake waterfront was entirely funded

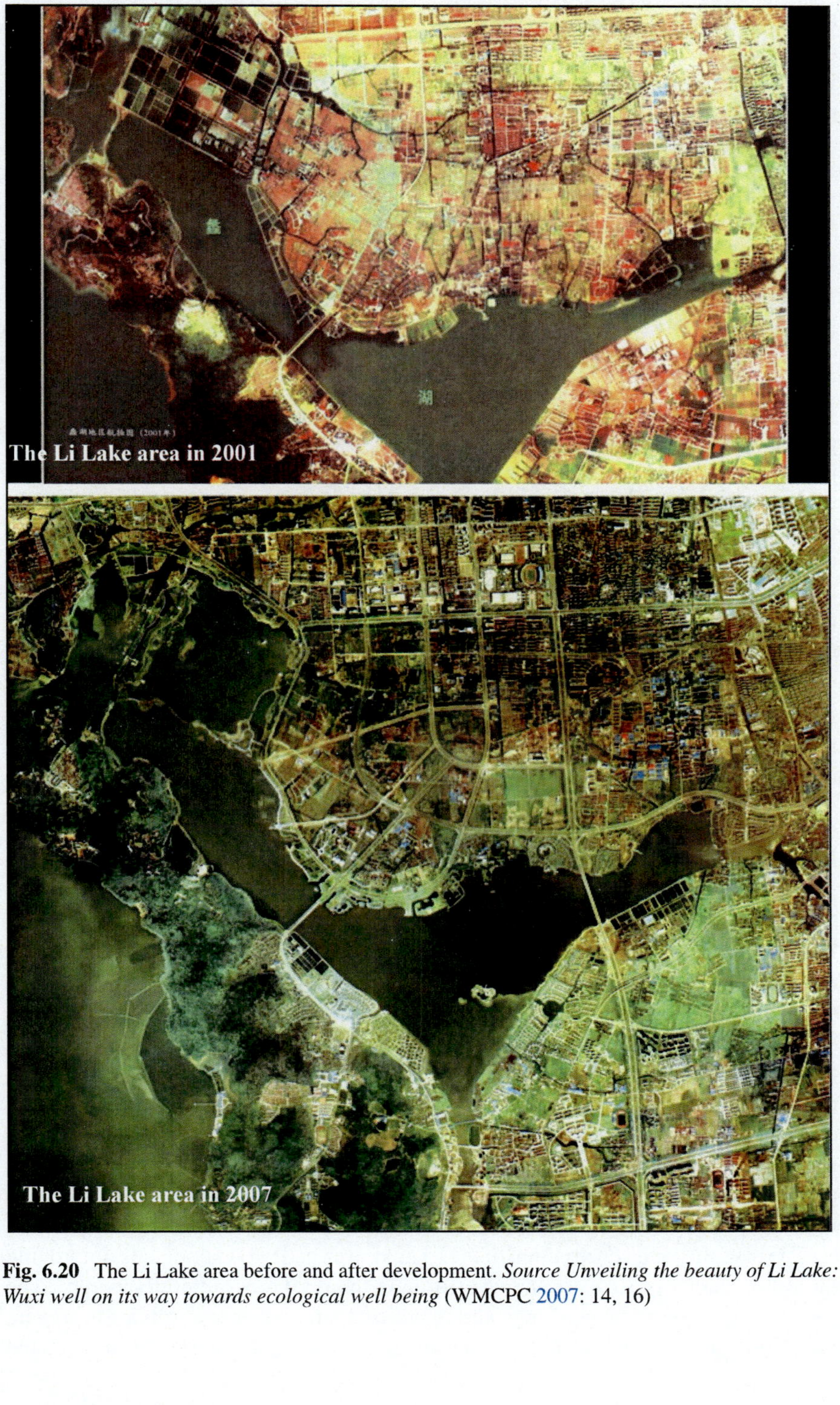

Fig. 6.20 The Li Lake area before and after development. *Source Unveiling the beauty of Li Lake: Wuxi well on its way towards ecological well being* (WMCPC 2007: 14, 16)

by the government. This fact, coupled with the greater involvement of the international design teams, indicates the value the government placed on the Western Li Lake waterfront and the consequent expectations of foreign landscape architectural services.

6.4 Conclusion

The local characteristics of the Li Lake area have been analysed in two levels: first, the economic and cultural significance of Jiang-nan, which forms the regional context of the Li Lake area; and second, the history of Wuxi and Li Lake, through which the Li Lake area is defined.

Regarding the regional context, the expression of place in Jiang-nan is underpinned by the water system of the Lake Tai plain and the historically water-associated production of food and activities. The expression of place was shown through the traditional spatial form of water towns and villages, the architectural features of buildings, gardens and bridges, and the customs and practices of everyday life. Such attributes constitute the everyday landscape of the Jiang-nan region, which is formed by people making a living using the natural conditions and resources of the region to produce their own sense of place. Within this sense of place, the traditional gardens and parks of Jiang-nan occupy a prominent position. The literary culture that gave rise to these gardens and their cultural meanings has also influenced Chinese culture as a whole. The regional sense of place has been highlighted by these gardens and parks and their iconic elements, which have in turn influenced the contemporary practice of creating places and landscapes in China.

Wuxi's industrial and commercial history and its water-associated production and trade brought a unique landscape that is characterised by water control and canals and a lasting urban form defined by canals. The Li Lake and Wuxi area have been also identified with traditional gardens and parks and the associated literary culture. Yet, the grand scale of Lake Tai, the folklore born from the myths of the great water body, and its mercantile culture all lend unique natural and cultural characteristics to the Li Lake area. Such uniqueness distinguishes Wuxi and the Li Lake area from the nearby water towns, cities and villages of Jiang-nan that are characterised by slender rivers and canals and small-scale inner lakes and private gardens. Therefore, the regional identity of the Li Lake area comprises two important aspects. First, it owes its identity to the scale and the wildness of its water body and the control and exploration of its water. Second, its identity arises from the iconic elements of the traditional gardens and parks of Jiang-nan and the associated literary culture. These two complementary and contrasting regional identities have persisted in the local memory and are reflected in various historical and contemporary records.

The planning and construction of the Li Lake area had two major goals: improving the water quality of Li Lake and transforming the rural areas into urban areas. The former was an environmental engineering project for the purpose of solving the severe pollution of Li Lake, which was also a result of its industrial history. The latter aimed at accommodating the urban population, providing waterfront parks for residents and improving Wuxi's image to promote its tourism economy. The environmental and economic focuses underpinned the requests made to landscape architectural firms from overseas. Specifically, there was an expectation that such firms would bring high-quality services related to land-use planning through landscape design and construction. The following chapters further analyse the processes and outcomes of international design practice between the beginning of 2002 and the end of 2005, when Australian landscape architects played key roles.

References

EDAW. (2002). Wuxi New Lake City, Master Plan Report.

Fang, P., & Qian, Y. (1999). *Classics of gardens: Appreciation of gardens and gardening in Jiangnan of China*. Hangzhou shi, Zhejiang ren min mei shu chu ban she.

Feng, X. (2007). Economists' view: Jiang-nan is wealth. *Chinese National Geography, 557*(3), 64–67.

Guo, H., & Li, Y. (2007). Habitant, picturesque scene and artistic conception: Landscape design of northern shore of Jincheng Bay. *Hua Zhong Architecture, 25,* 81–91.

Kahn, J. (2007, 5 Nov). In China, a Lake's Champion Imperials Himself. Retrieved 28 Apr 2008, from http://www.nytimes.com/2007/10/14/world/asia/14china.

Keswick, M. (1986). *The Chinese garden*. New York, London, St: Martin's Press; Academy Editions.

Lin, Y. (1936). *My country and my people*. London, W. Heinemann.

Pan, G. (2001). *The art of landscaping in Jiang-nan* Nanjing, Southeast University Press

Peng, Y. (1988). *Analysis on Chinese classical gradens*. Taibei shi, Di jing qi ye gu fen you xian gong si chu ban bu.

RDUDP. (2002). Concept Planning Design of Wuli Lake Area of Wuxi City.

Sun, Z., & Liu, J. et al. Eds. (2007). *Urban Science Research of Wuxi: Li Lake Special* Wuxi, Editorial board of Wuxi urban science research.

Sun, Z., & Shao, G. Eds. (2007). *Wuxi stepping toward an ecological water-mountain waterfront city: chronicles of the building of the new city of Li Lake* Wuxi, Wuxi's Municipal Committee of the Planning and Construction of Lihu City.

The Ministry of Environmental Protection of China. (2001, 25 Oct). The Tenth Five-year Plan of Lake Tai Pollution Prevention. Retrieved 28 Apr 2008, from http://www.mep.gov.cn/info/gw/huangfa/200110/t20011015_86475.htm.

Wei, D.-Y. (2006). A Chinese-English Dictionary (Revised Edition). English Department of Beijing Foreign Language University. Beijing, Foreign Language Teaching and Research Press.

Wuxi Statistic Bureau. (2007, Aug 21). "A survey of Wuxi." Retrieved Jul 21, 2008, from http://www.wxtj.gov.cn/sqjs/wxgk/index.shtml.

Wuxi Municipal Committee of the Planning and Construction of the Li Lake area, Ed. (2007). *Unveiling the beauty of Li Lake: Wuxi well on its way towards ecological well being* Phoenix Press Group and Phoenix Publishing House.

Yang, H. (1994). *A treatise on the garden of Jiangnan: a study into the classical art of landscape design of China*. Shanghai, Shanghai ren min chu ban she.

Yi, Y. (2008). Searching the history of Wuxi, ehancing the fascinating Wu culture. *Journal of Chinese Academy of Social Science*.

Zhang, J. (2007). Lake Tai and Jiang-nan. *Chinese National Geography, 557*(3), 128–138.

Zhou, W. (1990). *History of Chinese Classical Garden*. Beijing, Qing hua da xue chu ban she: Xin hua shu dian zong dian ke ji fa xing suo fa xing.

Zhuang, S. (1995). *Wuxi local history*. Nanjing, Jiangsu ren min chu ban she.

Chapter 7
The Practice of Australian Landscape Architects

Abstract This chapter explains the practices of the two international design teams involved in the project of Planning and Construction of the Li Lake Area from 2002 to 2005, when Australian landscape architects played key roles. This chapter also explains the composition of the two international teams and the ways in which they provided services. Chapter Seven has three major sections. Section One analyses the Wuxi Municipal Government's role as the key developer and its expectations of foreign landscape architectural firms. Section Two explains the composition and practice of the international design team that the clients recognized as the 'Australian landscape architectural firm'. Section Three explains the composition and practice of the international design team that the clients recognised as the 'American landscape architectural firm'.

Keywords Local expectations · Australian landscape architects · The RDUDP team · The EDAW team · Project phases

7.1 The Expectations of 'Foreign Landscape Architectural Firms'

The Wuxi Municipal government acted as the key developer for the planning and construction of the Li Lake area. Reasons for inviting landscape architectural firms from overseas by the government were implied in their developmental principles. As articulated in the previous chapter, one of the five principles was 'rigorously enforcing planning controls, opening the planning and design markets, and raising the threshold for granting construction permission' (Sun et al. 2007: 16). This principle was the first articulated by the government for developing the Li Lake area and aimed specifically at 'positioning the Li Lake area and its planning and design at a high standard and ensuring a high quality implementation' (2007: 16). This principle itself had four guidelines (2007: 16–17), showing the demand for foreign landscape architectural services.

The first guideline, 'rigorously enforcing planning control', was about the control of permission for land development by sub-developers (Sun et al. 2007: 16). It meant

Y. Zhang, *Place Making in International Practice of Landscape Architecture*,
Advanced Topics in Science and Technology in China 64,
https://doi.org/10.1007/978-981-16-2442-1_7

that sub-developers would be required to provide their proposals for further development, based upon the master plan of the Li Lake area, to win such permission (2007: 16). This plan was to be provided by the invited international design teams and approved by the government. In response to this guideline, in the first half of 2002, when the teams were preparing planning schemes for the Li Lake area, the government approval of land development to sub-developers had ceased temporarily (2007: 16). This control of land was also embodied in a government document issued in the second half of 2002, confirming the master plan provided would 'be used to guide the constructive detailed planning of the development of the [entire] Li Lake area' (2007: 8).

The second guideline was the 'opening of the planning and design market' (Sun et al. 2007: 16). This guideline was for the purposes of 'updating design principles, selecting design firms with high service standards, and reforming ways of design tendering' (2007: 16). Under such a guideline, two foreign landscape architectural firms were chosen 'after a careful investigation of a few world-influential environmental design companies and their principal designers' (2007: 16–17). The clients referred to the two firms as 'foreign landscape architectural firms': 'Australian Ron Dennis' and 'American EDAW Asia' (2007: 16–17). Although the method of team selection was not explained, the government clients' focus on the world's top landscape architectural firms was explained by the 2004 Wuxi governmental report, *Approaches and Experiences of Making a Model City of Wuxi*. In that report, EDAW Asia was chosen from those 'listed in the top ten world-famous design companies' and regarded as having 'successfully completed [the 2000] Sydney Olympic Village and Hong Kong Disney Land' (Wuxi Government 2004). The intention of updating local design ideas was also expressed by the local government, which is translated as follows (Sun et al. 2007: 8):

> Committing foreign design firms to plan the Li Lake area is beneficial, because it offers references for us of their planning experience, new concepts and theories. It helps to improve the planning standard of [our] city, the quality of the urban environment, and helps to construct an outstanding waterfront city.

The third guideline related to the reform of the local decision-making system (Sun et al. 2007: 17). This means that before governmental clients make decisions on plans provided by international design teams, they gather advice from the Wuxi residents and local experts. Under this guideline, plans of the Li Lake area were publicly exhibited, and residents of Wuxi were organized by the media to discuss the plans. Moreover, to 'tap the potential of the history and culture of Li Lake', and to 'plan the location and construction of cultural landscapes to enrich the cultural meaning of the Li Lake area' (2007: 17), authorities from cultural circles were invited to discuss the plans and offer advice on the names of roads, bridges, scenic spots and dykes (2007: 17). According to the 2002 news of Editorial Board of *China Wuxi* Journal (EBOCWJ) (2002) that was sourced from the Wuxi government website, on 20 September 2002 a Cultural Symposium for the Planning and Construction of the Li Lake Area was convened by the government to seek advice from cultural circles, including artists and poets. These authorities gave advice on the twenty-four scenic spots of the Li Lake area. Since that date, the government has held a Cultural Symposium on every eighth of January of the traditional Chinese calendar for the purpose of integrating local cultural values with the development of the Li Lake area (Sun et al. 2007: 34–35).

The fourth guideline concerned 'quality implementation' and referred to the reform of project management (Sun et al. 2007: 17). Such a reform was also aimed at 'learning from the experience of construction supervision of foreign countries' (2007: 17). In response to this guideline, 'a senior landscape architect of EDAW [the eventual design team] was invited to conduct rigorous construction supervision for the whole process, and consequently ensured the realisation of planning and design ideas' (2007: 17). This chief supervisor of the EDAW team, an Australian landscape architect, arrived at Wuxi on 8 September 2003 (Sun and Shao 2007: 53) and worked with the government for nearly half a year (Brookbank 2007: personal communication). The date of his arrival was recorded in the government's chronicle of the project's events, suggesting the value the clients placed on the construction phase by introducing construction supervision from overseas.

The four guidelines discussed above also describe the major functions of the planning and design provided by the international design teams: first, creative planning and design ideas that were distinguished from those provided by local planners and designers; second, a guiding plan to seek public opinion and integrate cultural meanings; third, a guiding plan to control further land development by sub-developers; and last, ensuring the construction standards and qualities of the built landscape.

The start of the environmental engineering component of the case project was marked by the establishment of the Wuxi Municipal Committee of Li Lake Comprehensive Water Treatment on 20 November 2001 (Sun et al. 2007: 146). The practical work of water treatment started about a half-year later on 22 May 2002, marked by the silt removal from the lake (2007: 22). The start of the urban development component of the case project was officially recorded as on 15 November 2002, when the government publicly announced the establishment of the 'Wuxi Municipal Committee of the Planning and Construction of the Li Lake Area' (WMCPC). The committee aimed at leading the planning and construction work (2007: 9). The team consisted of twenty-two heads of local governmental institutions, with the directors of the team being the deputy-executive mayor and the deputy-secretary generals of the Wuxi Municipal Government. The head of the Wuxi Planning Bureau (WPB) was top-listed amongst these team members (Sun et al. 2007: 9). Although the WMCPC was officially established on 15 November 2002, they started to contact the two international design teams nearly a year earlier, which was at the end of 2001 (Wulff 2007: personal communication; Harvison 2010: personal communication). Moreover, the formal commencement of the two international design teams was in early 2002, almost 10 months earlier before the formal establishment of WMCPC.

Although understood as an Australian firm and an American firm by the clients, both firms comprised several organisations that were geographically distinct, often at some distance. Therefore, in reality, these two firms were two international consortiums consisting of professionals drawn from across Australia and worldwide, respectively. Of these professionals, landscape architects from Australia played key roles in both international teams. This explains why the clients recognised the two teams as landscape architectural firms.

The 'Australian firm' participated in the planning phase for the entire 20 square kilometres of the Li Lake area from January to May 2002. The participation of the 'American firm' covered phases of planning the entire Li Lake area, then the landscape design of the northern shore of the Li Lake waterfront, the landscape design development of the northern shore of the Western Li Lake waterfront and its construction supervision. This spanned a four-year period from February 2002 to the end of 2005. The following sections further analyse the practices of the two international design teams, including their composition and the key role played by Australian landscape architects.

7.2 The Practice of the 'Australian Firm': The RDUDP Team

The full name of the Australian firm was *Ron Dennis Urban Design Planning and Business Consultancy* (RDUDP) (2002: front page). This consortium was an Australian-based team comprising consultants from different Australian-based firms. According to the planning report produced by the RDUDP team (2002: front page), the lead RDUDP consultant was geographically located in the Brisbane suburb of Albion. The report (2002: acknowledgements) also listed key members of the team in the following sequence: project director (Ron Dennis); director of master planning (Rodney Wulff); director of planning (Peter Crick); director of urban planning (Mark Doonar); director of transport (Neil Viney); director of Architecture (Herb Greenwood); marketing director (Frank); business director (Sarah); and interpreter (Eva and Cristal). (The last names of the last four participants were not provided in the report).

It is important to recognise that although Ron Dennis was the project director, it was the landscape architects from Tract Consultants who played key roles in the planning phase for the entire twenty square kilometres of the Li Lake area. The website of Tract Consultants (2008) showed that the directors of master planning (Wulff) and urban planning (Doonar) were principals of the Tract Melbourne and Brisbane offices, respectively. In addition, in the list of the 17th *Annual Awards of Excellence in Planning*, which was issued by the Planning Institute of Australia (PIA)—Queensland (QLD) Division (2003: 4), the RDUDP's plan for the Li Lake area received 'Certificates of Merit'. The entrants were listed in a sequence as follows: Tract Consultants, Ron Dennis Urban Design Planning and Business Consultancy, Perrott Lyon Mathieson Architects, and Wuxi Planning Bureau. The master planning director of the RDUDP team (Wulff 2007: personal communication) indicated that Tract did most of the work in this planning phase.

The major professional services provided by Tract Consultants suggest the key role played by Australian landscape architects in the RDUDP team. The firm Tract has the experience in providing landscape architectural services, including planning and design. According to its website (2008), Tract Consultants is a professional team that can 'operate beyond the usual scope expected of the landscape consultant and to contribute to large-scale, complex environmental and urban projects' through the 'bonding of landscape architects and planners'. The website (2008) also claimed that the expertise of the firm Tract was 'built on uniting two professional disciplines [planning and landscape design], that developed in isolation for over half a century'. The director of master planning (Wulff 2007: personal communication) indicated that the cooperation between Tract Consultants and Ron Dennis was based upon professional links, as they already had business relationships in Australia on other projects in Australia and China. The director (Wulff 2007: personal communication) also indicated that the Melbourne office of Tract Consultants was involved mostly in the practice team.

The planning report prepared for WPB by the RDUDP team indicates that the team commenced work in January 2002. The report shows that the team was invited to 'prepare concept plans and designs for improvements to Wuli [Li] Lake and its environs' (RDUDP 2002: 8). In this report, a section titled 'the brief' suggests how the government clients expected them to work (2002: 8):

> The work concentrated on the layout, forms and functional aspects of the land uses and facilities around the lake, as well as landscape features, water activities, and external influences. The latter include the planned university town to the south of the study area. The boundaries of the study area were seen as flexible and where the project team thought it appropriate, consideration was given to the incorporation of areas beyond these boundaries.

Among the above major tasks, the team was requested specifically to cover certain aspects of work, including (RDUDP 2002: 8):

- Clearly expressing an overall plan and design intent for the study area
- Clearly defining the functions of differing land use areas
- Delineating the form and type of public spaces and visual features of the specific sites around the lake
- Defining a traffic network (and a hierarchy of roads, and networks for bicycle paths, tourist roads, and ferry routes), along with conceptual designs for main roads
- Describing the controls of building forms and the characteristic of spaces via master plans and perspective drawings
- Suggestions about the functional areas and development controls to be incorporated in the future urban planning of the study area
- Recommendations about which precincts need to be the focus of the next stage of the project.

The clients requested that the outcomes of the above tasks should be presented in 'large size plans' (RDUDP 2002: 8) and via three conceptual structure plans; a final conceptual structure plan; a final concept master plan; bird's-eye view perspectives; an overall location plan; a present status plan; analytical drawings including topography, traffic, urban spaces and parks, external influences, and precincts;

objectives/design principles, sections of typical roads, open spaces and water-edge treatments and a land-use plan.

To comply with the clients' requests, the RDUDP team spent nearly five months completing the work of the planning phase, from January to May 2002 (RDUDP 2002: cover page, 8). The director of master planning (Wulff 2007: personal communication) indicated that the team was given only ten days for the development of the initial planning concepts. A further two days were spent sketching these concepts. These sketches, dated 18 January, 2002, were presented to the Wuxi Municipal government. This means that the team had less than 18 days from when its work officially commenced to the presentation of the preliminary concept structure plans, which also included a few analytical plans. Each sketch of the three structure plans was a large hand drawing (Fig. 7.1).

After presenting the three conceptual structure plans, the RDUDP team developed them into a final structure plan and presented the draft final plan to the Wuxi Municipal government in March 2002. After receiving feedback on the draft final structure plan in April 2002 from the clients, the team left Wuxi and developed the final master

Fig. 7.1 The practice of the RDUDP team—meetings with clients and local experts. *Source* Top—RDUDP's planning report (2002: 7); bottom—*Wuxi stepping toward an ecological water-mountain waterfront city* (Sun and Shao 2007: 22)

Table 7.1 The practice of the RDUDP team—milestones. *Source* Adapted from RDUDP's planning report for the Li Lake area (2002)

Time	Milestones
2002 January	Work commenced; three preliminary structure plans produced; major presentations to government of the three plans on 18/01/2002
February	Regional analysis for final structural plan produced, 25/02/2002
March	Feedback on regional analysis was received from clients, the final concept structure plan produced and presented to clients
April	Meeting with local experts for the draft report: *Concept Planning Design of Wuli* [Li] *Lake Area of Wuxi City,* 19/04/2002
May	The completion of the final report: *Concept Planning Design of Wuli* [Li] *Lake Area of Wuxi City*, 07/05/2002

plan in Australia and produced the planning report: *Concept Planning Design of Wuli* [Li] *Lake Area of Wuxi City*. Dated May 07, 2002, this planning report also recorded briefly the process and outcomes of the RDUDP team's work. Table 7.1 shows the milestones in the RDUDP team's work for the planning phase of the development of the Li Lake area.

The RDUDP team described their methods of conducting consultancy work for the case project as 'the often used intensive workshop process' (RDUDP 2002: 9). According to the planning report, this intensive workshop lasted nearly four months, from 18 January to 19 April 2002. During this time, the team studied the site, produced three conceptual structure plans, developed a final conceptual structure plan, and drafted a planning report. The team then returned to Australia to prepare the final planning report. During this process, the team incorporated various activities to gain local understanding and to gather information essential for their consultancy. According to the planning report, these included (2002: 9):

1. Visiting Wuxi city and inspecting the Li Lake area, including guided and self-guided tours, assisted by WPB.
2. Receiving advice from local experts such as civil engineers.
3. Developing planning and design responses according to the brief, and requesting new information during the information gathering stage, including:

    (1) Presentation of the refined brief.
    (2) Briefing by senior government officials, receiving and answering questions of fact, policy and emphasis.
    (3) Meeting with government officers representing water transport, public facility and infrastructure, environmental protection, water conservation, tourism, parks and garden, culture and heritage and urban planning.

4. Conducting a series of formal and informal presentations to and receiving detailed feedback from the Wuxi Municipal Government, including senior government officials and representatives, government officers and experts.

The above activities suggest that the period for producing three large conceptual structure plans, which was less than 18 days, was the most intensive stage of the entire process. During this short time, the RDUDP team visited Wuxi, the Li Lake area, received advice from local experts, responded to the brief, refined the brief, met senior government officials, and finally produced large sketches comprising three preliminary structure plans. This, however, suggests that the three structure plans were based upon local knowledge gained in a very short period. This further suggests that such local knowledge was either readily available in a written format or visually easily recognisable. Moreover, the above activities for gaining local knowledge were dominated by engagement between the RDUDP team and the various actors of the Wuxi Municipal government. Of the latter, WPB took a major role in terms of inviting the team and organizing meetings with governmental representatives. The director of master planning (Wulff 2010: personal communication) indicated that in the intensive workshop period, a WPB minister came almost every night to take questions from the RDUDP team and then brought these questions to local experts and returned with the latter's suggestions.

In parallel with the RDUDP team's work, the government clients also developed guidelines for the project. The RDUDP's planning report shows that during the team's visits of the Li Lake area and discussions with the government representatives, 'the Wuxi [municipal] city government has [had] determined that the city will focus on tourism as one of the main economic drivers' because the new arrangement for holidays and weekends in China since 2000 provided opportunities to develop the local tourism economy (RDUDP 2002: 10). The planning report of the RDUDP team indicates that in the local area there were no strategies or research on this perceived dramatic change in tourism at the time (2002: 10). This goal for tourism, however, was not clearly expressed in the brief cited in the planning report. According to the report, after the RDUDP team presented three conceptual structure plans, the government 'determined which aspects would be preferred elements of a Final Structure Plan' (2002: 34). These aspects included that 'Wuli [Li] lake is to serve the local residents as well as tourists'; and 'open space around the lake is to be available for local residents as well as tourists ' (2002: 34).

In February 2002, when the clients' decision for a tourism focus was increasingly clear, the RDUDP team emphasised their concerns for both Wuxi residents and tourists in terms of the content of their three structure plans. After the three structure plans were produced, the clients accepted RDUDP's proposal to consider both local residents and tourists as important contributions. This decision was recorded in RDUDP's planning report as made by the government clients in April 2002. This concern was retained in the stage of producing the planning report of the 'Final concept[ual] Master Plan' (RDUDP 2002: 40), which was completed in May 2002. In this planning report, the phrase 'local residents' was always put ahead of the word 'tourists'.

After the RDUDP team completed their planning report in May 2002, it ceased working on the Li Lake area project. The master planning director (Wulff 2010: personal communication) indicated that he did not receive any further response from the clients after that time. He also (Wulff 2010: personal communication) expressed his regret for this lack of communication, although the team was well-paid. In September 2002, however, again through Ron Dennis, the RDUDP team soon commenced planning work for an area about 20 kilometres away from the Li Lake area. Also located on the waterfront of Lake Tai, the area is called Double Bay. It is important to recognise that about fifty kilometres away, on the outskirts of Suzhou, Jing-ji Lake is located. Another international design team, the so-called EDAW Asia, had provided planning services for the Jing-ji Lake project, which started in 1998. Figure 7.2 shows the proximity of the locations (coloured in red) of the three projects in which Australian landscape architects played key roles.

The planning report for the Double Bay area, completed in June 2003, had a structure similar to the report prepared for the Li Lake area. Table 7.2 shows the milestones in the work of the RDUDP team for their second project of the waterfront of Lake Tai. It also shows that outcomes and stages were similar to those provided for the planning of the Li Lake area, including three structure plans, a final structure plan, and a master plan. However, the clients of the Double Bay area remained unknown to the key player of the RDUDP team, Tract Consultants. The front page of the planning report states 'clients unknown'. This further suggests a communication gap between the key providers of planning and design services of the RDUDP team and the key decision makers of the clients' side of the two projects.

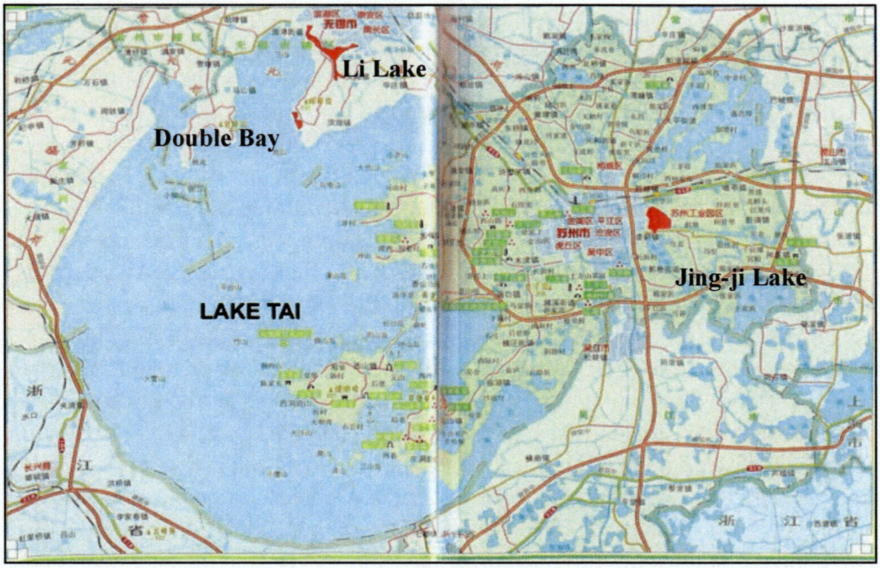

Fig. 7.2 Proximity of Li Lake, Double Bay and Jing-ji Lake. *Source Jiangshu Map* (Yu 2001: 44–45), notation by author

Table 7.2 The practice of the RDUDP team for the Double Bay area—milestones. *Source* Adapted from RDUDP's planning report for the Double Bay area (2003)

Time	Milestones
September 2002	Second project Double Bay commences, three structure plans produced
October 2002	On 18, submission of preferred structure plan for Double Bay On 30, completion of final preferred structure plan for Double Bay
November 2002	On the 29, completion of the 'Landscape Master Plan' report
December 2002	Early Dec, submission of 'Landscape Master Plan' report
April 2003	Formal presentation of the Landscape Master Plan for Double Bay
May 2003	A series of comments received after the presentation to clients
June 2003	On 6, completion of the 'Double Bay, Tai Hu: analysis, concept options, preferred structure plan' report. On 24, report updated

7.3 The Practice of the 'American Firm': The EDAW Team

Similar to the composition of the RDUDP team, the 'American firm' consisted of staff from the various offices of the international firm EDAW. It is important to recognise that although the clients referred to this consortium as 'American EDAW Asia' (Sun et al. 2007: 16–17), Australian landscape architects from the EDAW Brisbane office played a key role in the practice, especially between the end of 2001 and 2005. During this time, the international firm EDAW (including EDAW Asia) experienced a reorganisation. In 2004, when the EDAW team had completed the landscape design development of the western Li Lake waterfront, the name EDAW Asia no longer existed and the firm's name had reverted to EDAW.[1] In December 2005, EDAW became part of the AECOM,[2] a global provider of professional technical and management support services. The website of the AECOM claims that it is formed from many of the world's finest engineering, design, environmental and planning companies.[3] The leadership of the EDAW offices in the EDAW team also shifted along with the reorganisation.

The practice of the team for the Li Lake area involved four of the worldwide offices of the international firm EDAW, including those in San Francisco, Hong Kong, Brisbane, and (later) Shanghai. Of these four offices, it was the EDAW Brisbane office staff and the Australian landscape architects who played a key role in providing the planning, landscape design, landscape design development and construction supervision services. This is evident in the e-news of EDAW dated 20 March 2002 and issued by the EDAW San Francisco office. Winsor (2002) of the EDAW San Francisco office said in the e-news that the composition of the team was an 'interdisciplinary collaboration of HK [Hong Kong], BRIS [Brisbane], and SF [San Francisco] offices'. He (2002) explained:

[1] http://www.edaw.com.hk/cm/whoWeAre/history/history-asia.htm, retrieved 08 July 2008.
[2] http://www.edaw.com/whoWeAre/history/history50.aspx, retrieved 08 July 2008.
[3] http://www.aecom.com/About/The+AECOM+Brand, retrieved 06 September 2010.

> A collaborative, international EDAW team, that includes the Hong Kong, Brisbane and San Francisco offices, has been working with the city of Wuxi, China (PRC) to create a major new master plan that is an excellent example of a DEEP infusion project.
>
> The team members include Andross Chan (HK) as principle-in-charge and Les Wood (BRIS) as project manager and senior designer, assisted by Tony Dickson (BRIS) as lead urban planner for the New Lake City [the Li Lake area] Master Plan.
>
> Mark Winsor (SF) leads the environmental strategy program and Mike Conlon (SF) leads the economic development program. Other team participants in Brisbane include Szczepan Urbanowski, urban design and architecture; John Wood, recreation planning [consultant]; Michael Erickson, planning oversight; and David Walker, graphics. The Shanghai office staff provides additional assistance.

The above explanation also shows that in addition to the EDAW Brisbane staff, a range of professionals in Brisbane, Australia, participated in the work of planning and designing the Li Lake area. These Brisbane professionals included those subcontracted by the EDAW Brisbane office, such as Szczepan Urbanowski, who provided work for urban design and architecture. Urbanowski was also a key contributing consultant, as Urbanowski was listed as one of the two applicants for receiving the international award bestowed to the Li Lake project, which was issued by the Urban Land Institute (ULI) (ULI Website 2006).

But while the e-news mentioned above states that the Shanghai office staff provided additional assistance, an e-news issued by EDAW on January 30, 2003, suggests that by mid-2002 the EDAW Shanghai office had taken leadership of the team (Chu 2003). Chu (2003) of the EDAW Hong Kong office claimed:

> In mid 2002, under the leadership of EDAW-Shanghai, the Shanghai and Brisbane offices completed a winning scheme of the 20-square kilometre Wuxi New Lake City [the Li Lake area]; a mixed-use lake community comprising of civic, residential and natural environments.
>
> Clearly demonstrating the fundamentals of EDAW's "DEEP" practice, the project involved staff from the Shanghai, Brisbane and San Francisco offices.

Chu (2003) also claimed that the winning planning scheme produced in mid-2002 was 'a milestone for EDAW, essentially bringing a flow of follow-on work with an estimated contract value of US $2 millions'. This included, first, landscape concept to design development of a 6 km greenbelt and public waterfront area; second, landscape concept to design development of 100 ha commercial centre and central green; and third, landscape concept and implementation of all street furniture and signage throughout the Li Lake waterfront.

A comparison of these two e-news items suggests that project leadership had shifted from the EDAW Hong Kong office in March to the Shanghai office by mid-2002. In contrast, from the time that work commenced in February 2002, to the creation of 'a major new master plan' in March 2002, as reported by Winsor (2002), and then to the completion of 'a winning scheme' in mid-2002, as reported by Chu (2003), the Brisbane office was continuously acting as a key participant in the project. According to its principal, the EDAW Brisbane office actually received the project brief more directly through the EDAW Hong Kong office and 'then prepared the technical response', while 'the Hong Kong office would have negotiated the terms and cost' (Harvison 2010: personal communication). This suggests that the Australian

Table 7.3 The practice of the EDAW team—contributions of EDAW Brisbane

Principal planning and design documents	EDAW Offices	Date
1. New Lake City, Wuxi, China: Master Plan Presentation	Brisbane	Mid 2002
2. Wuxi New Lake City: Master Plan Report	Brisbane	August 2002
3. Five Mile Lake Landscape Design	Brisbane	December 2002
4. Landscape Design Development Package: Wuxi Lakefront	Brisbane	July 2003
5. Landscape Design Development Package: Fishing Dragon Island	Brisbane	November 2003

Source The case study

landscape architects of EDAW Brisbane's office took responsibility for professional matters from the beginning, while the Hong Kong office took responsibility for business matters.

The major role played by Australian landscape architects in the EDAW team is also evident in the key planning and design documents that were produced. Table 7.3 shows five key sets of planning, landscape design and landscape design development documents. The principal confirmed that the EDAW Brisbane office prepared them all (Harvison 2010: personal communication). These planning and design documents covered the three phases of the work (Fig. 6.12): first, planning of the entire 20 square-kilometre Li Lake area; second, landscape design of the northern shore waterfront of Li Lake; and last, landscape design development of the northern shore waterfront of Western Li Lake.

The planning report and master plan presentation package both suggest that the EDAW team proposed New Lake City as the name of the Li Lake area in the future. This differs from the name used by the RDUDP team in their planning report, which kept the original name of the Li Lake area. The proposed name, however, was similar to that finally used by the clients (Li Lake New City), suggesting that the view of the EDAW team coincided with that of the clients.

The planning report produced in August 2002 also confirmed the lead role of the EDAW Brisbane office in the planning phase, saying that the report was 'prepared by a collaborative team of EDAW from offices in Australia, China and America' (EDAW 2002 August: 2). After 8 September 2003, an Australian landscape architect of the EDAW team acted as construction supervisor for the construction work of the waterfront of Western Li Lake. He spent 'almost ½ [half] year living and working alongside the [Wuxi] government and began to understand their city' and also acted as guide for the clients on their visit to Brisbane (Brookbank 2007: personal communication). This Australian landscape architect, Brookbank, became a key member of the EDAW Shanghai office and continued working on the project after the completion of construction of the Western Li Lake waterfront.

Given that the principal planning and design work was provided by the EDAW Brisbane office, and also that the highly praised construction supervision work was provided by the Australian landscape architect, Australian landscape architects in the EDAW team can be seen to have made major contributions through all project phases, from conceptualizing the Li Lake area to assuring the quality of materializing planning and design concepts on the waterfront. This contrasts with the clients' designation of the EDAW team as an 'American firm'.

Although Australian landscape architects played a critical role in the EDAW team, the often-emphasized 'DEEP' approach (Winsor 2002; Chu 2003) also suggests that the practice of the EDAW team was worldwide and interdisciplinary. 'DEEP' is EDAW's trademark approach for planning and design; it stands for integration of Design, Environment, Economics and Planning (Lau 2004). The work components dealing with environment and economics were provided by the EDAW San Francisco office. In the early stage of producing three conceptual structure plans, the San Francisco office provided eight environmental strategies and economic programs for the Li Lake area (Winsor 2002). This also suggests that, generally, the EDAW Brisbane office dealt with the work of 'DP' (Design and Planning), the San Francisco office with the 'EE' (Environment and Economics), the Shanghai office with the construction, and the Hong Kong office with business issues.

The fact that the Hong Kong office dealt with business issues shows the use of the ethnic Chinese area of Hong Kong as a stepping stone to bridge different business cultures. In contrast, the primary conceptual work regarding DEEP was provided within the relatively similar cultures of the consortium of the EDAW team, in America and Australia. It could be argued that EDAW's approach to these Chinese projects differs from its approaches to projects in western countries or South-eastern Asia, where as Fuller argues (Nevill 2006) a wall of legislation provides a relatively regular procedure of planning and design practices. Using an ethnic Chinese area as a stepping stone suggests that personal relationships in business plays an initial and also crucial role in design and planning competitions. This aligns with the research of APSFADTRC and Forshaw (1996) that was discussed in Sect. 4.3.

Officially, the team commenced work in February 2002 (EDAW 2002 August: 2), nearly a month later than the RDUDP team did. The principal of the EDAW Brisbane office (Harvison 2010: personal communication) recalled that the team received the project brief 'as part of the Request for [a] Proposal'; also, 'as part of the tender[,] a brief and scope of work would have been included' and the brief had been refined through coordination meetings after commission. In the section of *Introduction and Background* of EDAW's New Lake City planning report, the brief was summarised as follows (EDAW 2002 August: 2):

> The City required a guiding Master Plan to develop the study area into an integrated tourism, recreation and residential centre to provide impetus for economic growth, greater residential amenity and a wider profile for Wuxi as a destination city.
>
> Key goals of the clients were to: increase tourism, strengthen the economic base of the city, and beautify the city and its lake setting.

The above requests by the government clients suggest their strong expectation for planning and design solutions that would improve the tourism economy and that by this time these expectations were expressed more explicitly to EDAW than they had been to the RDUDP team. The principal of the EDAW Brisbane office (Harvison 2007: personal communication) also indicated that the message received from the clients about the planning of the Li Lake area emphasised tourists rather than local residents.

The EDAW team spent nearly six months in the planning phases, which lasted from February to August 2002. The fact that the RDUDP team finished the planning report in late May 2002 and the EDAW team announced their winning scheme in mid-2002 (Chu 2003) suggests that the clients' decision to proceed with the master plan provided by the EDAW team was made in mid-2002. In fact, such a decision was made by WPB, as WPB recommended EDAW's plan to the government. This is evident in an official document issued by the Wuxi Municipal Government to WPB on 20 December 2002. In this document, the government confirmed WPB's selection of EDAW's master plan (Sun et al. 2007: 8). In the document, the government also suggested the following guide for further development of the selected master plan, translated as follows (2007: 8):

> The following work of landscape design of the waterfront of Li Lake and the design of important landscape nodes should be based upon the master plan provided by the EDAW team as a basis, and should integrate the merits of the other plan [that is provided by the RDUDP team].

The date of issue of this official document suggests the development of the selected master plan commenced after 20 December 2002. In reality, the amendment to the master plan started soon after EDAW's plan was selected in mid-2002, marked by the completion of their New Lake City master plan report in August 2002. In September 2002, there was also a public exhibition (Fig. 7.3) of the selected master plan for the purpose of seeking advice from Wuxi residents and local cultural circles (EBOCWJ 2002; Sun et al. 2007). This suggests that, from mid-2002, some desirable features in the planning scheme proposed by the RDUDP team had been integrated in that of the EDAW team.

Moreover, the landscape design of the northern shore waterfront of Li Lake was completed by the EDAW team in December 2002. This was followed by the landscape design development of the western part and its parallel construction in 2003 when the EDAW team had its own construction supervision on site. Table 7.4 shows the milestones in the work of the EDAW team from the beginning of 2002 to the end of 2005 and the participation of EDAW's worldwide offices.

According to EDAW's planning report, the process the EDAW team employed to carry out the planning phase had three stages: 'site investigation', 'concept master planning options' and 'final master plan' (EDAW 2002 August: 2). The report also shows that in the planning phase the approach of the EDAW team to gaining local knowledge included site inspection, client workshops, presentation and client feedback (EDAW 2002 August: 2). Although EDAW's planning report introduced their participatory approach quite briefly, the outlined approaches are quite similar to those of the RDUDP team. First, both teams used stages of information-gathering through self-guided study and meeting with governmental representatives and local experts. Second, WPB acted as a key representative of government clients, inviting and providing support for self-study and organizing meetings for both teams. The key role of WPB in WMCPC was evident by the fact that WPB recommended a

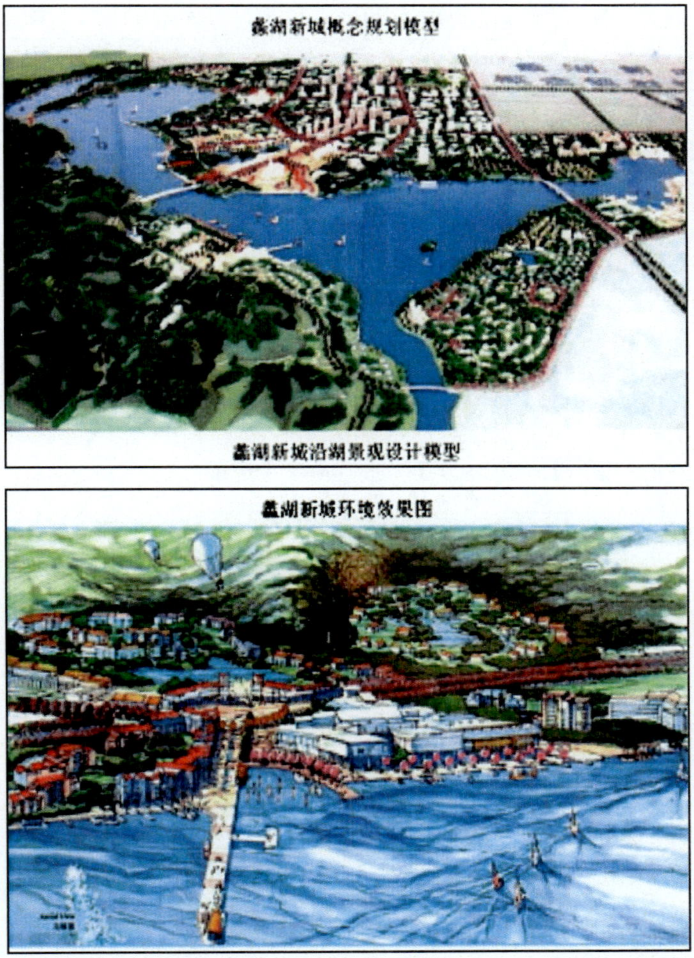

Fig. 7.3 The image for the exhibition of EDAW's planning scheme. *Source* Editorial Board of *China Wuxi* Journal (2002). Top: the model of Conceptual Master Plan. Bottom: the pespective drawing of the New City

preferred master plan to the Wuxi Municipal Government. Third, both international teams achieved direct engagement with the government clients through formal meetings and informal workshops, while meetings appeared to be dominant methods of communication. The following figure illustrates the dominant method of engagement between the EDAW team and the top local stakeholder, the Wuxi Municipal government: the formal meeting.

In addition to similar methods of engaging with the clients and sites, a further comparison of the processes and timelines of the RDUDP and EDAW teams reveals that the two international design teams experienced similar procedures and time allowed in the planning stage. The only obvious difference is the participation

Table 7.4 The practice of the EDAW team—milestones

Time	Milestones	EDAW Offices
February 2002	Work commenced	Hong Kong, Brisbane, San Francisco
11 March 2002	Complete three concept structure plans, presenting to the mayor and city staff of Wuxi	
Mid 2002		
Pre-August	Complete *New Lake City, Wuxi, China—Master Plan Presentation*	Shanghai, Brisbane, San Francisco
August	Complete *Wuxi New Lake City—Master Plan Report*	
December 2002	Complete *Five Mile* [Li] *Lake Landscape Design*	Shanghai Brisbane
16 July 2003	Complete *Landscape Design Development Package: Wuxi Lakefront stage 2*, about 16.7 ha	
8 September 2003	Australian Landscape Architect, Brookbank, arrived at Wuxi, acting as an on-the-spot construction supervisor	
28 November 2003	Completing *Landscape Design Development package*: *Wuxi Li Lake—Fishing Dragon Island*, about 7 ha	
1 May 2004	More park lands opened, including the above island	Shanghai
30 April 2005	Bo-gong Ecological Island opened	
30 September 2005	Li Lake Central Park opened	
31 December 2005	The construction of Li Dyke complete, marking the completion of the waterfront of Western Li Lake	

Source Adapted by author

sequence with that of the EDAW team beginning about one month later than that of RDUDP. Before 25 February, when the RDUDP team was going to produce the draft of the final structure plan, the EDAW team had in all likelihood started work. When the RDUDP team formally presented the final structure plan to the Wuxi Municipal Government in March 2002, the EDAW team could just have presented their three structure plans to the 'mayor and the city staff' (Winsor 2002). Also, during the time the RDUDP team returned to Australia to complete the final planning report of 'Concept Planning Design of Wuli [Li] Lake Area of Wuxi City, stage 1 report' in May 2002, the EDAW team also nearly completed their final plan (in 'mid-2002') (Chu 2003) in Brisbane (Harvison 2007). It could be argued that the work with RDUDP better prepared the clients for the later involvement of the EDAW team. This may have facilitated the clients' communication with the EDAW team, as they had experienced all the procedures with the RDUDP team beforehand. This, however, exposes a problem with regard to the matter of fairness in client dealing between competing planning and design consultancies.

Compared with the RDUDP team, the EDAW team appears to have been more ready to adapt to this operational difference, as EDAW had already been involved in the Jing-ji Lake project in the nearby city of Suzhou (Fig. 7.4) three years before, where the Brisbane office staff were also key players (Harvison 2007: personal communication). Furthermore, the practices of the two international design teams suggest that the key planners of the RDUDP team had a less direct engagement with the decision maker of WMCPC. It has become apparent during this research that the RDUDP team were not aware of the participation of the EDAW Brisbane office until the completed Western Li Lake waterfront received international awards for EDAW (Wulff 2007: personal communication) at the end of 2005. By contrast, the EDAW team announced their winning scheme in mid-2002, indicating their awareness of the competition. The partnership of the EDAW team with the local government was further enhanced in the phases after planning through the transformation of leadership from the EDAW Hong Kong office to the Shanghai office in the period since late 2002, the time when the EDAW Brisbane office developed the master plan and provided the landscape design of the northern shore of the Western waterfront.

The landscape design package of the Li Lake waterfront was provided by the end of 2002, four months after provision of the planning report in August 2002. During this time, leadership of the EDAW team shifted from the Hong Kong office to the Shanghai office. The distance between the clients and the EDAW team was

Fig. 7.4 The practice of the EDAW team—meetings with clients and local experts. *Source Wuxi stepping toward an ecological water-mountain waterfront city* (Sun and Shao 2007: 7)

greatly reduced. Although there is no specific indication as to how EDAW provided landscape design services, the landscape design package shows wider references to local history. This is further analysed in Chapter Eight. It could be argued that the enhanced interaction between the clients and the EDAW team through the EDAW Shanghai office facilitated the further work of the EDAW Brisbane office. This is evidenced by an Australian landscape architect being posted in Wuxi for a long-term purpose as construction supervisor in 2003, when the EDAW Brisbane office was providing landscape design development of Western Li Lake waterfront and at the same time the site was under intensive construction.

The participatory approach of the EDAW team changed greatly after 2003. The director of WMCPC (Yang 2007: personal communication) recalled that an Australian landscape architect carried out design directly in a Wuxi restaurant. The Australian landscape architect (Brookbank 2007: personal communication) described his construction supervision this way: 'the process for guiding construction was not formal construction management but rather using the expertise of the company [EDAW] to influence the client [to] manage its own construction'. Since 2004, the time when EDAW Asia no longer existed and the name of the firm reverted to EDAW, the EDAW team gradually transformed its working focus for the project from Brisbane to Shanghai. Construction supervision by the Australian landscape architect proceeded, and he became a senior staff member of the EDAW Shanghai office. Also, other senior staff with a Chinese background began to join the team, such as Zhou (Kalamaros 2006). All of these suggest that a closer interaction between the Australian landscape architects of the EDAW team and local stakeholders occurred in the year of 2003, together with the localisation of the practice.

7.4 Conclusion

The practices of the two international design teams in the development of the Li Lake area present three levels of complexity: first, an organisational complexity, referring to geographically distant companies or branch offices for both teams; second, a disciplinary complexity, referring to types of services covering land-use planning, environmental strategies, economic programs, landscape design, architectural design and construction supervision; and lastly, an operational complexity, referring to the overlapping procedures of the government approval of master planning and the development from master planning to landscape design, and the parallel landscape design development and construction.

Given that the two international design teams were recognized as landscape architectural firms by the government clients of Wuxi, the profession of landscape architecture can be seen as significant in the development of the Li Lake area. This suggests a more prominent role of landscape architects working internationally in dealing with lands being affected by environmental degradation and urban expansion in China. The principal contribution of Australian landscape architects can be seen through their provision of services in regard to planning, landscape design, landscape design development and construction supervision, especially landscape architects from the Tract Melbourne office and the EDAW Brisbane office. Other professionals based in Brisbane also provided a range of consultancy services under the contract of the EDAW Brisbane office, suggesting the potential influence of Australian professionals on the physical outcomes of the planning and design of the Li Lake Area.

The EDAW team continued into the landscape design phase before government approval of their master plan, as the landscape design package was completed at about same time as the issue of the approval. This fact suggests that operation of practice on-the-ground may very likely have differed from the procedures operating in the practitioners' own countries and thus required a more flexible approach. The communication gap between the key planners of the RDUDP team and WMCPC also suggests the significance of a direct association with clients, even and especially when they are culturally and geographically distant from the consultants.

Finally, in the planning practice of the two international design teams, the clients' intention to integrate local history and culture in the development was increasingly articulated. This is shown in the practical work of the clients involving Wuxi residents and local experts to comment on international design teams' master planning. The next chapter further analyses the outcomes throughout the planning and design phases and the local characteristics that were interpreted in the principal planning and design documents, as well as the constructed waterfront.

References

Australia Parliament Senate Foreign Affairs Defence and Trade References Committee and M. Forshaw (1996). *Australia China relations.* Canberra, Parliament of the Commonwealth of Australia.

Brookbank, P. (2007). Interview with Yun Zhang on 06 November at EDAW Shanghai office.

Chu, A. (2003, 30 Jan). Wuxi New Lake City: Win generates a flood of downstream work. Retrieved 3 Oct, 2007, from http://ed.edaw.com/firm/fwCommunications/eNews/eNewsArticle.aspx?aId=32359.

EDAW. (2002). Wuxi New Lake City, Master Plan Report.

Editorial Board of *China Wuxi* Journal. (2002). "Brief introduction of detailed planning of Li Lake New City." Retrieved May 7, 2008, from http://www.wuxi.gov.cn/intowx/survey/admin/20069/20069527768.shtml.

Harvison, G. (2007). Interview with Yun Zhang on 3 October at EDAW Brisbane' office.

Harvison, G. (2010). Communication through email, sent to University of Melbourne, on 25 May 2010.

Kalamaros, A. (2006, 28 July). Wuxi client accepts UIL award in Tokyo. Retrieved 3 Oct, 2007, from http://ed.edaw.com/firm/fwCommunications/eNews/eNewsArticle.aspx?aId=72822.

Lau, P. (2004). EDAW: the Integrated World." *Chinese Landscape Architecture*, (1):12–21.

Nevill, J. (2006). Look both ways. Retrieved 3 Oct, 2007, from http://propertycouncil.gravitymax.com.au/nat/page.asp?622=281749&e_page=17330&=....

Planning Institute Australia—Queensland division. (2003). 17th Annual awards of excellence in planning. Retrieved May 06, 2010, from http://www.planning.org.au.

RDUDP. (2002). Concept Planning Design of Wuli Lake Area of Wuxi City.

RDUDP. (2003). Double Bay, Tai Hu: Analysis, Concept Options, and Preferred Structure Plan.

Sun, Z., & Shao, G. Eds. (2007). *Wuxi stepping toward an ecological water-mountain waterfront city: chronicles of the building of the new city of Li Lake* Wuxi, Wuxi's Municipal Committee of the Planning and Construction of Lihu City.

Sun, Z., & Liu, J. et al., Eds. (2007). *Urban Science Research of Wuxi: Li Lake Special* Wuxi, Editorial board of Wuxi urban science research.

Tract Website. (2008). Tract: landscape architects, urban designers, town planners. Retrieved 21 May 2008, from http://www.tract.com.au.

ULI Website. (2006). ULI Award winner—Wuxi Li Lake parklands: Project summary. Retrieved Oct 3, 2007, from http://www.casestudies.uli.org/profile.aspx?j=7599&p=5&c=22.

Winsor, M. (2002, 20 Mar). DEEP Fusion Project in Wuxi, China: Interdisciplinary collaboration of HK, BRIS, SF offices. Retrieved 3 Oct, 2007, from http://ed.edaw.com/firm/fwCommunications/eNews/eNewsArticle.aspx?aId=24187.

Wulff, R. (2007). Interview with Yun Zhang on 25 September, at Tract consultants' Melbourne office.

Wulff, R. (2010). Communication through telephone, called to University of Melbourne, on 12 May 2010.

Wuxi Government. (2004, Aug 30). Approaches and Experiences of Making a Model City of Wuxi. Retrieved Jun 20, 2008, from http://big5.mep.gov.cn/gate/big5/www.mep.gov.cn/cont/mhcity/mfcsxx/zyzfjy/200408/t20040830_60884.htm.

Yang, Z.-Y. (2007). Conversation with Yun Zhang on 27 December at the office of Wuxi's Municipal Committee of the Planning and Construction of Lihu City.

Yu, Y.-P. (2001). The map volume of Jiangsu Province, Mapping Press of Chengdu.

Chapter 8
The Identified Local Characteristics

Check for updates

Abstract Chapter Eight explains the phase outcomes of the practice of two international design teams, focusing on how the characteristics of Wuxi and Jiang-nan were interpreted in their principal planning and design documents. This chapter also analyses the representation of these characteristics in the designed and constructed Western Li Lake waterfront and reasons for the waterfront receiving international awards and Chinese national awards. The chapter has four major sections. Section One analyses the principal documents of the planning of the Li Lake area, in which both international design teams were involved. Section Two analyses the principal documents of the landscape design of the northern shore of the Li Lake waterfront, in which one of the two international teams was further involved. Section Three analyses the principal documents of the landscape design development of the Western Li Lake waterfront. Section Four analyses the constructed Western Li Lake waterfront through a discussion of how the waterfront area has been used by the public and referred to in the announcements of international and Chinese national awards.

Keywords Master planning · Urban design · Waterfronts · Local fishery culture · Architectural forms · Local activities

8.1 The Planning Outcomes

In the planning phase, the two international design teams (the RDUDP team and the EDAW team) provided three conceptual structure plan options, and a master plan developed from these options. Also, each team provided a planning report that included analyses of the studied area and planning solutions explained by plans, sections and texts. The following paragraphs analyse how local characteristics were interpreted in these plans, drawings and texts.

Y. Zhang, *Place Making in International Practice of Landscape Architecture*, Advanced Topics in Science and Technology in China 64,
https://doi.org/10.1007/978-981-16-2442-1_8

8.1.1 The RDUDP Team

Of the RDUDP team, traces of local characteristics of the Li Lake area were shown in the 'Analysis' section in their planning report (RDUDP 2002: 11). The report listed 'specific aspects' of the existing Li Lake area in six categories, the first of which was the overall location of the study area and its relation to other aspects of the Wuxi city area; second, the important urban spaces, such as parks; third, the topographical features; fourth, view corridors; fifth, the basic traffic system; and sixth, external influences affecting the study area.

In the above categories, the specific aspects discussed included the city-wide context, such as the proximity of Wuxi to Shanghai and Nanjing; important scenic spots and tourist destinations around Li Lake and Lake Tai; bio-physical features of the land and water; the main canals and the ancient Grand Canal; the original fish farms as a source of water pollution; significant infrastructure, such as flood gates and dams; and the network of existing transportation (RDUDP 2002: 11–30). These specific aspects suggest two approaches to knowing the area for the team. One was receiving and studying background information from the clients. The other was the investigation and study of the site. The director of master planning (Wulff 2007: personal communication) also indicated that the clients provided plenty of background information about the environment and history of the Li Lake area and the region, similar to those received when conducting work in Australia.

The listed of important urban spaces and parks in the report (RDUDP 2002) included not only the existing entities of the site, but also future entities that had been planned. The report also suggests additional attention paid by the team beyond those requested by the project brief, including the following: first, important boulevard planting—'the elegant informal planting of camphor laurel trees'; and second, 'culturally and historically significant sites, significant trees, significant public infrastructure, such as the flood gate and Du Shan Dam, and the new bridge planned to be built over the eastern part of the lake' (RDUDP 2002: 13) (Fig. 8.1, top). The analysis of view corridors also suggests the RDUDP team's extensive on-site visits. These include not only the existing important vistas and view corridors but also the identified potential viewing spots (Fig. 8.1, bottom). The director (Wulff 2007, 2010: personal communication) of master planning indicated that the team conducted extensive studies of the site and Wuxi city, saying they identified what they believed to be important local features, such as doll factories and the Three Islands, which were not included in the background information provided by the clients.

Based upon the above specific aspects, the team proposed eight function zones (precincts) of the Li Lake area (Fig. 8.2), which differed from the six precincts proposed later by the EDAW team. These precincts were proposed for the purposes of maximising environmental sustainability, tourism, infrastructure, and internal water (canal) opportunities. According to the report, they were also proposed for creating real estate opportunities, the conductivities of Li Lake to surrounding areas, facilitating water-dependent land use, enhancing scenic quality, providing public accessibility, and 'serv[ing] local residents as well as tourists' (RDUDP 2002).

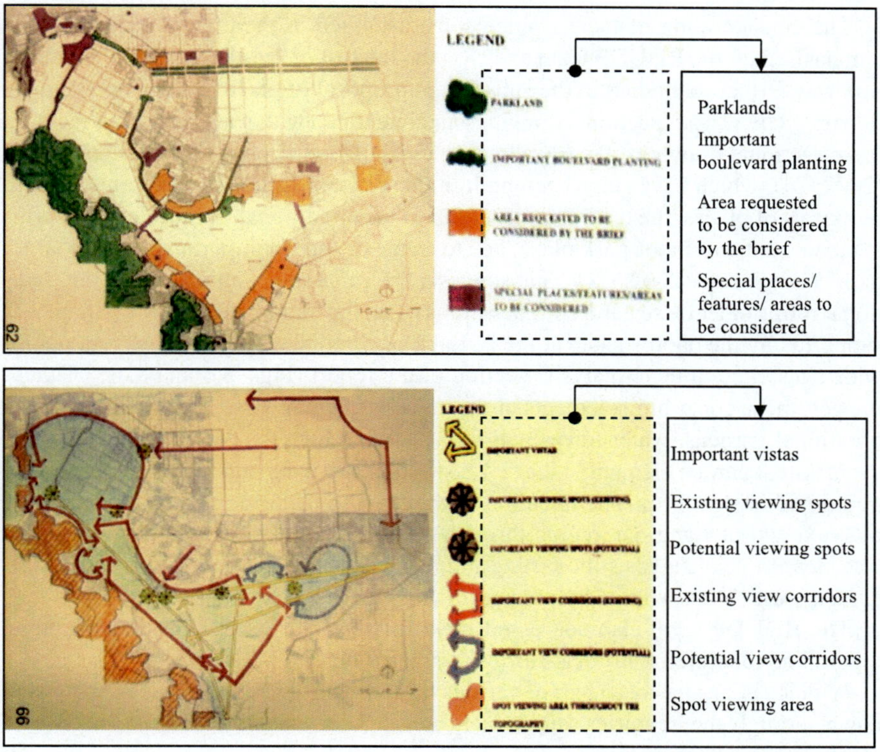

Fig. 8.1 RDUDP's analysis of urban space, parks and view corridors. *Source* The planning report of the RDUDP team (2002: 62, 63), notation by author

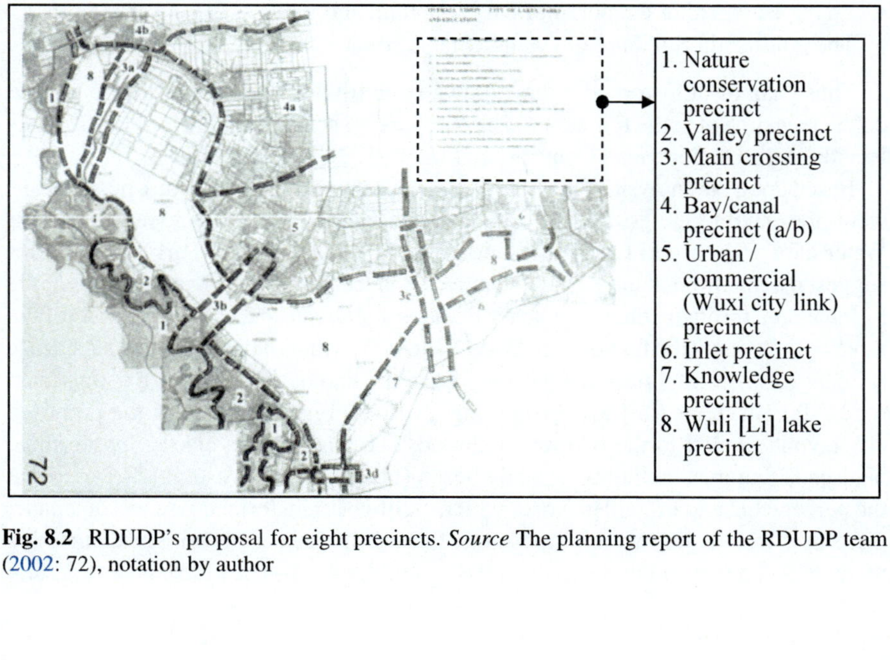

Fig. 8.2 RDUDP's proposal for eight precincts. *Source* The planning report of the RDUDP team (2002: 72), notation by author

The consideration of local residents, as discussed in the previous chapter, was emphasised by the RDUDP team and won the interest of the clients. This consideration was further embodied in the subsequent proposal of places for public activities illustrated by various sectional drawings and integrated with water edge treatment. In these sectional drawings, boardwalks and a waterfront amphitheatre were proposed (2002: 24), which have since been built in the Western Li Lake waterfront. In these proposals of places, the team also proposed to 'use traditional Chinese garden forms to create contrasts in the park/plaza, and to create or signify important nodes in larger parks and plazas' (2002: 23). This suggests the value the team placed on the traditional forms of Chinese gardens. The emphasis on the traditional Chinese garden was supported by the photos taken at the Li Lake area and the thirty images from worldwide sites cited in the 'Analysis' section of the report (Figs. 8.3 and 8.4). Of these images, those containing elements of the traditional architectural forms of Jiang-nan traditional gardens, water towns and villages were highlighted, such as the use of the classical garden elements—Lake Tai stones for rockery and moon doors, and the traditional bridge form. The intention of the team to incorporate the characteristics of Wuxi and of Jiang-nan region through the use of traditional architectural forms was clear. It also suggests the lasting influence on the West from Chinese classical gardens, which are typified by gardens of Jiang-nan.

The RDUDP team also addressed local folklore and legends in their planning report, and identified four local stories as significant (RDUDP 2002: 5):

Wuli [Li] Lake has been part of various legends and stories that have been handed down through the centuries, and which now further enhance the Lake's sense of place. These stories include -

- Zhuang Zhu firing fish in the fish village
- Fan Yi [Li] and Xi shi boating on the Xio [xiao] Wu Lake
- Zhang Bo opening the porch of Du Mountain, and
- The goldfish in Shi Tang of Chang Guang Stream

These names of historical figures drawn from folklore and legends were used to define the name of some functional areas proposed, such as 'Xi Shi Harbor', 'Lover's Islands' and 'Dragon Head Conventional Inlet' (RDUDP 2002: 48).

Based upon the analysis of the Li Lake area, three conceptual structure plan options were presented to the government on 18 January 2002, which were entitled 'Wuxi Water City', 'Wuli [Li] Lake Park' and 'Wuli [Li] Lake Central Park'. These titles suggest the intention of the RDUDP team to enhance the natural characteristics of the Li Lake area through enhancing water features and green space. Comparisons of the three options show the first option, Wuxi Water City, offered the largest water surface (Fig. 8.5). A second influence on the layout of the first option was two existing main roads: Tai Hu [Lake Tai] Road and Qing Qi Road. Figure 8.5 shows the proposed city layout parallel to the two retained roads. The first option also proposed three landscape features, including: a sandy beach (Fig. 8.5: 1) in Western Li Lake and at the perceivable end of Tai Hu Road, which was to be transformed into a promenade; three new islands that were proposed in Eastern Li Lake with bridges connecting them (Fig. 8.5: 2); and new canals in the central lands of the Li Lake area (Fig. 8.5:

Fig. 8.3 RDUDP's citation of photos of the Li Lake area. *Source* The planning report of the RDUDP team (2002: 4, 11), with marking as originally on the report copy supplied to the researcher

3). The sandy beach and islands have since been built in western and eastern Li Lake, respectively. The location of the built islands is the same as that proposed in this first option.

The second option, Wuxi Lake Park, provided more parkland by reclaiming lands from existing fish ponds (Fig. 8.6). Similar to the first option, the two existing roads were retained. However, the two roads became the primary influence determining the layout of the whole plan, including a main axis crossing their junction. Buildings

Fig. 8.4 RDUDP's citation of images across the world. *Source* The planning report of the RDUDP team (2002: 21), with marking as originally on the report copy supplied to the researcher

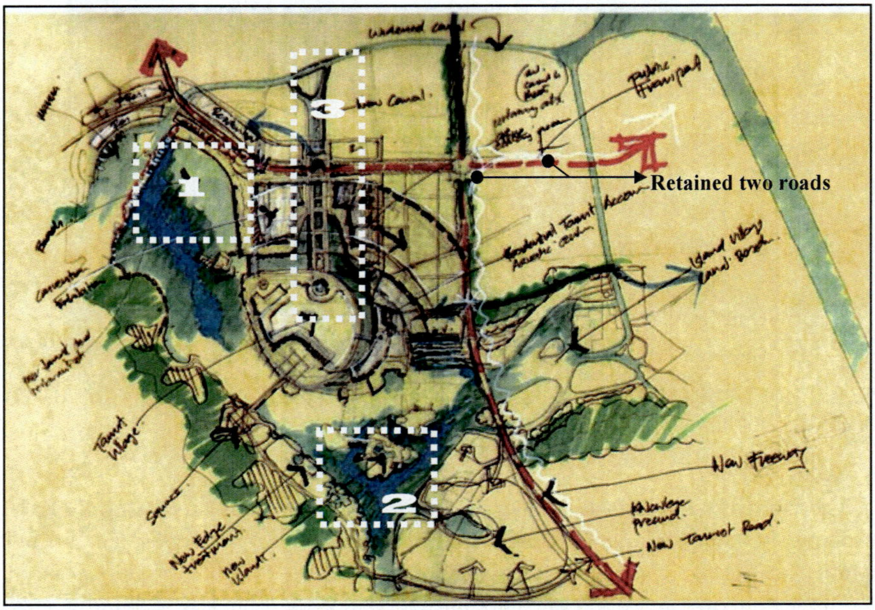

Fig. 8.5 RDUDP's conceptual structure plan, option one. *Source* The planning report of the RDUDP team (2002: 85), notation by author

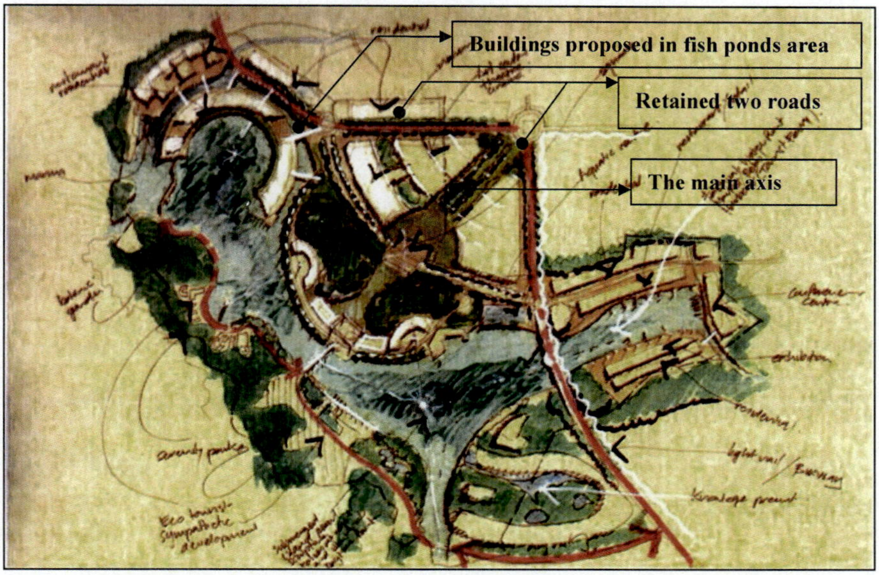

Fig. 8.6 RDUDP's conceptual structure plan, option two. *Source* The planning report of the RDUDP team (2002: 90), notations by author

and parklands were deployed along this axis. Thus, a strong geometrical pattern was formed and further enhanced by circular buildings deployed in the west.

The third option, Wuli [Li] Lake Central Park, also generated more land for development by reclaiming land from existing fish ponds, and reducing the lake surface area (Fig. 8.7). Similar to the second option, the two existing roads determined the layout of the plan and formed a plaza at their junction, forming the basis for a more symmetrical deployment of parks, buildings and plazas. The repetition of a circular pattern throughout the central park and cultural centre enhanced its more formalised structure. Similar to the first option, the third option also proposed a beach integrated with residential development and an island in Eastern Li Lake as a site for water sports.

All three conceptual structure plans suggest a dominating centripetal pattern. Also, islands and a sandy beach were proposed in two of the three options. These two ideas, together with the canals and extending the water of Li Lake into the central land as proposed in the first option, were carried forward to the master plan (Fig. 8.8: 1, 2, 3, 4). This suggests the value the RDUDP team placed on the four ideas and on the first option, as the four ideas were all embodied in the first option. The director of master planning (Wulff 2007: personal communication) also indicated his preference for the first option.

After presenting the three options to the governmental clients of Wuxi, the RDUDP team was asked by the clients to integrate the merits of the three options into a master plan (RDUDP 2002; Wulff 2007: personal communication). As was described in the planning report, the master plan had five key features (RDUDP 2002: 39)

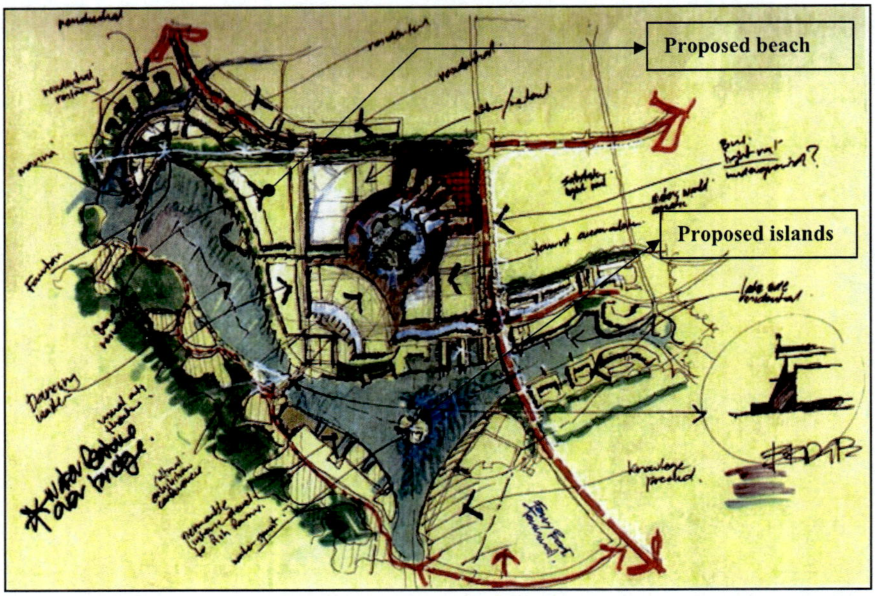

Fig. 8.7 RDUDP's conceptual structure plan, option three. *Source* The planning report of the RDUDP team (2002: 96), notations by author

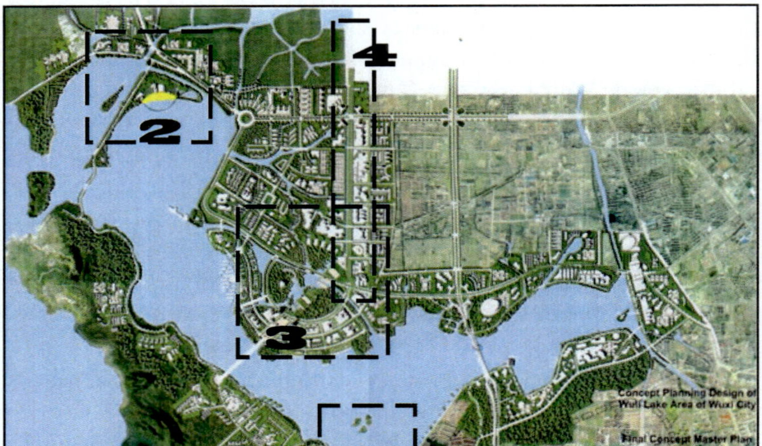

Fig. 8.8 RDUDP's final master plan. *Source* The planning report of the RDUDP team (2002: 103), notations by author

including first, balancing organic and formal patterns; second, lake surface area and new canals; third, functional areas and the road network; fourth, open spaces and points of focus; and fifth, broad building control envelopes. These key features were illustrated by a land use plan, a transport plan and perspective sketches of potential development. Also, extensive explanatory texts were included for sixteen proposed functional areas (RDUDP 2002: 30–56). Overall, the planning report is more explanatory in comparison with the more pictorial planning report provided by the EDAW team.

The layout of the master plan was again dominated by a circular pattern (Fig. 8.8). The team claimed that this master plan drew 'heavily on the organic nature of good city form, one which has enabled the city of Wuxi to prosper for 3000 years' and was a balance of 'organic and formal patterns' (RDUDP 2002: 39). It could be argued that the circular pattern preferred by the team is related to the ancient urban form of Wuxi city, the quasi-lozenge urban form as was described in Chapter Six (Fig. 6.2–3). Such a circular urban form had been recorded and lasted for over 500 years and was characterised by three features including first, the city being encircled by canals; second, canals crossing at the central part of the city; and third, the walled city. It could be argued that these three features were captured in RDUDP's master plan, represented through the deployment of canals around the circular central parklands, canals crossing the centre, and buildings surrounding the circular parklands.

Although the four ideas of islands, a beach, canals and water extension (Fig. 8.8: 1, 2, 3, 4) were proposed in the master plan in locations similar to the first conceptual structure plan option, their sizes were reduced. This enabled a more connected central land area and a continuous water surface for the lake, as has been characterised in the second and third options. The three proposed islands also add spatial layers to what had been a continuous water surface. The director of master planning also indicated that the team identified 'Three Islands' as an important theme of their plan. As was explained in Chapter Seven, the existing Li Lake is separated from Lake Tai by Turtle Head Peninsula at west. Close to the other side of the peninsula, there are also islands called 'Three Islands'. Moreover, the term 'Three Islands' is associated with Taoism in Chinese philosophy (Keswick 1986).

From their analysis of the study of the area to their provisions of a master plan, the intention to integrate local characteristics was evident in the manner in which the RDUDP team identified important planting, historical, and cultural sites beyond the requirements in the brief. Their proposals for using the architectural forms of traditional gardens, water towns, and the long-lasting urban forms of Wuxi all suggest this intention. It was also evident in their proposal of 'Three Islands' and canals in the Li Lake area, and in the incorporation of the names of historical personages for their proposed functional areas. Finally, it was evident in their emphasis on considerations of waterfront usage for Wuxi local residents, as well as tourists. Their proposal of a sandy beach, boardwalk, amphitheatre, natural conservation precinct in Western Li Lake and islands in Eastern Li Lake all have subsequently been built in the present-day Li Lake area, despite the team not having any further involvement in the following work.

In fact, the Chinese clients requested the EDAW team to integrate what they considered the merits of the master plan produced by the RDUDP team. Such a request however, exposes a problem with regard to intellectual property in planning and design. While integrating ideas of competing design teams is not uncommon in local Chinese planning and design competitions, it is not common in the West or considered fair practice. It could be argued that the rooted correlative Chinese thinking, as discussed in Sect. 2.3.2, may lead to clients' habit of combining different ideas, thus to stressing the importance of outcomes as a whole whilst disregarding the specific values of individual contributions.

8.1.2 The EDAW Team

For the EDAW team, traces of local characteristics of the Li Lake area were shown in the section 'Context, Existing Conditions and Options' of their planning report (EDAW 2002 August: v). The EDAW team provided a 'Comparison [of Wuxi] to Other Major Waterfront Cities', and analysed its 'national, regional and local context' (2002 August: 4) and that of the Li Lake area. The comparison was drawn with the cities of Sydney, Brisbane, London, and Shanghai, because (2002 August: 6):

> The study area can be compared to other international cities that have experienced revitalization of their waterfront areas undertaken over the past decades with considerable success…These cities are all relevant to Wuxi as they are all major centres of rich history and culture that have identified the need to renew their inner city areas. These cities have also traditionally 'turned their back' on their best assets—their rivers and waterfront areas which comprise mostly industrial land uses—to create rich, vibrant mixed-use urban areas with economic benefits for the whole city.

Of the above four cities, each was analysed for its four critical factors of success in waterfront renovation. These include (2002 August: 7):

Sydney, Australia

- Tourism and image focused on the harbor and waterways
- Iconic and distinctive waterfront architecture
- Promotion of entertainment and vibrant mixed[-]use districts (e.g. Darling Harbor)
- Extensive use of public transport system on the waterways

London, England

- Focus on traditional wharf and maritime areas for urban renewal and theming
- Provision of extensive new living and community area[s]
- Contemporary waterfront landscape treatments used
- Provision of light rail vehicle (LRV) system to facilitate access and circulation

Brisbane, Australia

- Continuous waterfront public access
- Major parklands along waterfront corridor
- Creation of new urban village areas with differing character and land uses
- Incorporation of river-based rapid transit system (City-Cat)

Shanghai, China

- Blend of older historical and new elements
- Focused [focuses] on the city's rivers and creeks
- Use of iconic architecture
- Extensive public transport system

These waterfront cities are all modern and western (or highly globalised, such as Shanghai). This suggests the intention by the team to look to the future rather than to ancient China and the physical traditional architectural forms.

Similar to the RDUDP team, EDAW's (2002 August: 8–10) analysis of national, regional and local context addressed Wuxi's location in China, its proximity to Shanghai, its thousands-of-year-old history and civilisation, its tourism economy and natural environment. In its analysis of regional context, attention was clearly and explicitly paid to tourism (2002 August: 9):

> The region of Wuxi is strongly identified as Wu culture. Key economic industries of the region are agriculture, industry and manufacturing. It is also one of the top ten tourism nodes in China. Tourism was first introduced in the early 1900's and is dominated by day trippers. The annual visitation is approximately 16 million visitors per annum. Most tourism visitation is domestic with 30% of these from the region and 20% nationally from other parts of China. International tourism is not high with approximately 400,000 international visitors per annum.

The EDAW team referred the city-wide context to the 'local context' (2002 August: 4). Their analysis addressed not only Wuxi's 4000-year settlement and 2000 years of cultural history, but also the tourism economy, the area's climate, and Li Lake's proximity to Wuxi's CBD (2002 August: 10). Figure 8.9 shows two plans illustrating the regional context (top) and city framework (bottom) presented in EDAW's master planning report. The former suggests a transition from nature to semi-nature and urban along with the areas from Lake Tai to Li Lake and Wuxi's CBD. The latter determined the Li Lake waterfronts as a 'tourism lake edge and tourism frame' (EDAW 2002 August: 10). The focus on tourism is clearly emphasised in EDAW's master planning report.

Of the analysis of the Li Lake area itself, the report listed three major 'elements', including the '5 Mile [Li] Lake edge', 'Turtle Head Island [Islet]', and 'The Canal System'. This suggests the identification of the historic landscape around the site. The team also provided four plans regarding 'Existing and Proposed Development', 'Existing Tourism and Recreation Node' (Fig. 8.10), 'Road Hierarchy' and 'Key Views' (Fig. 8.11). Except for the focus on tourism, most of the analytical plans of

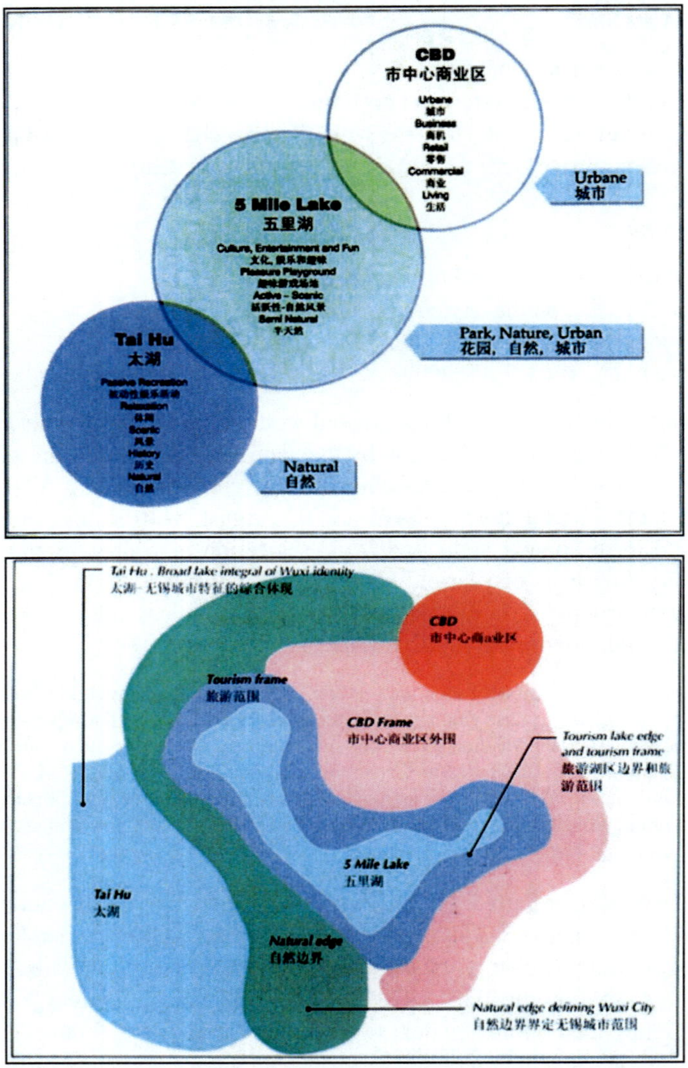

Fig. 8.9 EDAW's analysis of the regional and city-wide context. *Source* The master planning report of the EDAW team (2002 August: 9–10): *Wuxi New Lake City, Master Plan Report*

the EDAW team have titles similar to those used by the RDUDP team, suggesting that the primary request of clients from the two teams were similar.

The team also developed planning principles with regard to 'Land Use Planning', 'Tourism', 'the Natural Environment', 'the Built Environment', 'Access and Circulation', and 'Open Space'. The focus on tourism was again clearly meant in the explanatory texts of 'Land Use Planning' and 'Tourism'. Of 'the Natural Environment', it suggests the need in development to: 'protect and enhance the aesthetic

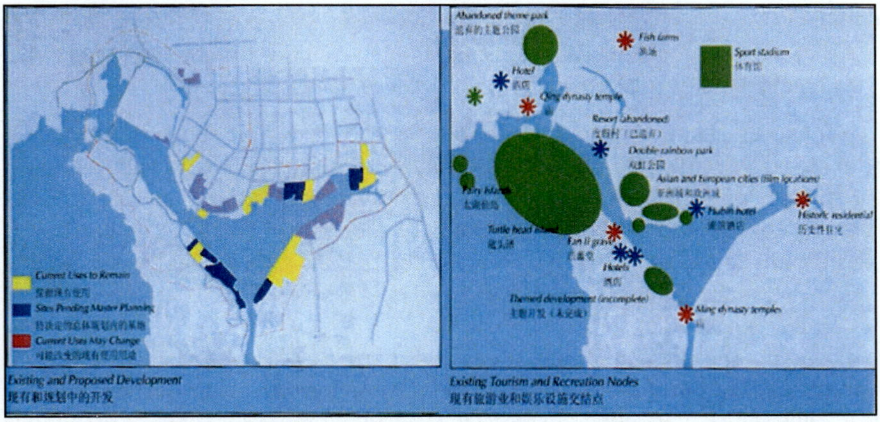

Fig. 8.10 EDAW's analysis—Development and Tourism. *Source* The master planning report of the EDAW team (2002 August: 13)

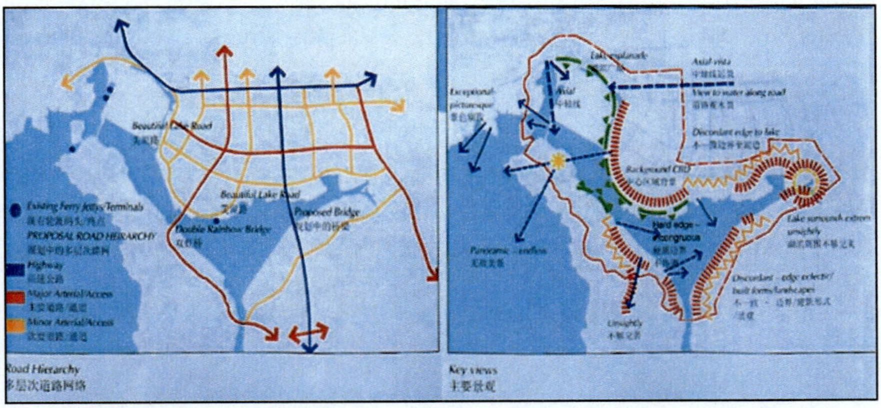

Fig. 8.11 EDAW's analysis—Road Hierarchy and Key Views. *Source* The master planning report of the EDAW team (2002 August: 13)

landscape qualities of 5 Mile [Li] Lake and its surrounding environs that contribute to the distinctive character of the area' (EDAW 2002 August: 27). For 'the built environment', there were suggestions to (2002 August: 27–28):

- Create an attractive and memorable city image for visitors.
- Achieve a built form that highlights the basic structural elements of the City's development pattern providing clarity of City image and enhancing the attractiveness of the urban landscape.
- Identify and protect places and objects of recognized cultural and heritage significance.

- Establish and enhance the local urban character promoting a distinctive local identity and sense of place and providing opportunities for community activity, interaction and participation.

For 'Access and Circulation' and 'Open Space', there were suggestions again to promote an attractive and memorable city image. These planning principles suggest the intention of the EDAW team to integrate both natural and cultural characteristics of the Li Lake area, promoting a new image and identity of the city. Compared with the apparent emphasis on image building of the Li Lake area, it could be argued that the concern for community activities is relatively vague, as ways to address this were not further explained.

Following the above, the EDAW team also provided three conceptual structure plan options, which were presented to the government clients on 11 March 2002, nearly one and a half months after the RDUDP team. Comparisons of the three options show that differences existed mainly in the central area proposed for the new city. Also, all options employed a radial pattern to control the structure. Contrasting with the generally similar radial pattern, however, each plan offered a different solution for the water body in the central area, including its form, size, and the integration of water with the road system. The top-listed conceptual structure option shows an ellipse-shaped centre to the city (Fig. 8.12) with water bodies at each corner of the centre, and a radiating road system. The main axis was determined by the junction

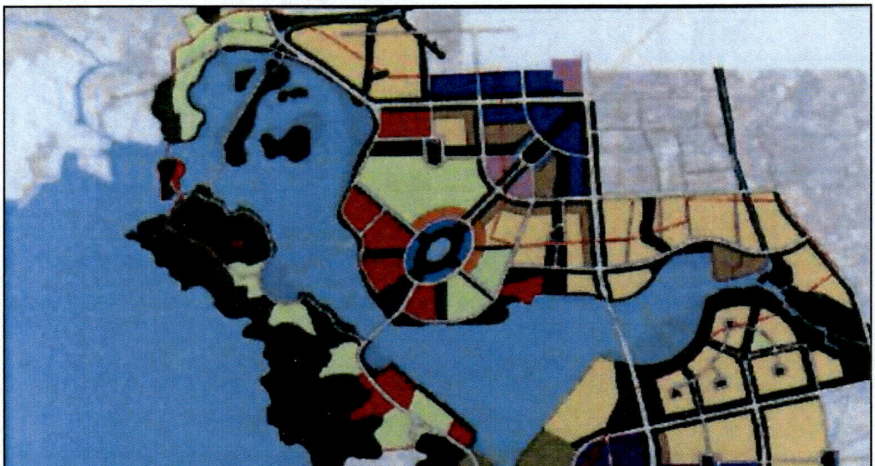

Fig. 8.12 EDAW's conceptual structure plan, option one. *Source* The master plan presentation package of the EDAW team (2002 pre-August: 2): *New Lake City, Wuxi China, Master Plan Presentation*

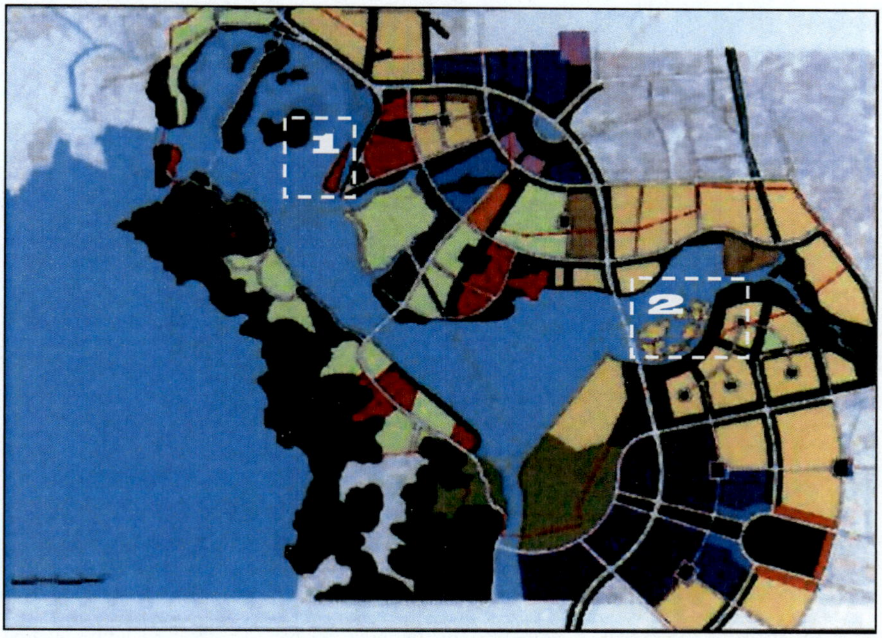

Fig. 8.13 EDAW's conceptual structure plan, option two. *Source* The master plan presentation package of the EDAW team (2002 pre-August: 2), notation by author

of the two retained roads and the central point of the ellipse. Compared with other options, however, the central water body in this option is the smallest.

In the second option (Fig. 8.13), the two existing roads were modified and joined as a single circular road. This circular road provided a pattern opening towards Li Lake and contrasted with the enclosed structure characterised by the ellipse form in the first option. The structure of sub-roads and the trapezoidal water bodies in the central area were coherent with such a pattern, opening towards Li Lake in one direction and converging at the crossing-point of the newly joined roads in the other. At this junction, a semi-circular plaza was deployed, emphasising the central point of the pattern. Other parts of this plan were similar to the first option, except for Western Li Lake, where a major island was added along the waterfront (Fig. 8.13: 1) in the location of the present Fisher Father's Islet. Also, in Eastern Li Lake, a few small islands were proposed and wetlands were enlarged (Fig. 8.13: 2).

In the third option (Fig. 8.14), the major pattern is a hybrid of the previous two options. The northern part of the city area (Fig. 8.14: 1) was inherited from the first option, a structure determined by the two existing roads and the enclosed ellipse at the centre. The southern part of the city area (Fig. 8.14: 2) was inherited from the second option, opening towards Li Lake. Again, other parts of the third structure plan were similar to the second option, except for Eastern Li Lake, where the built area increased and the shape of the water body became more geometric (Fig. 8.14: 3, 4).

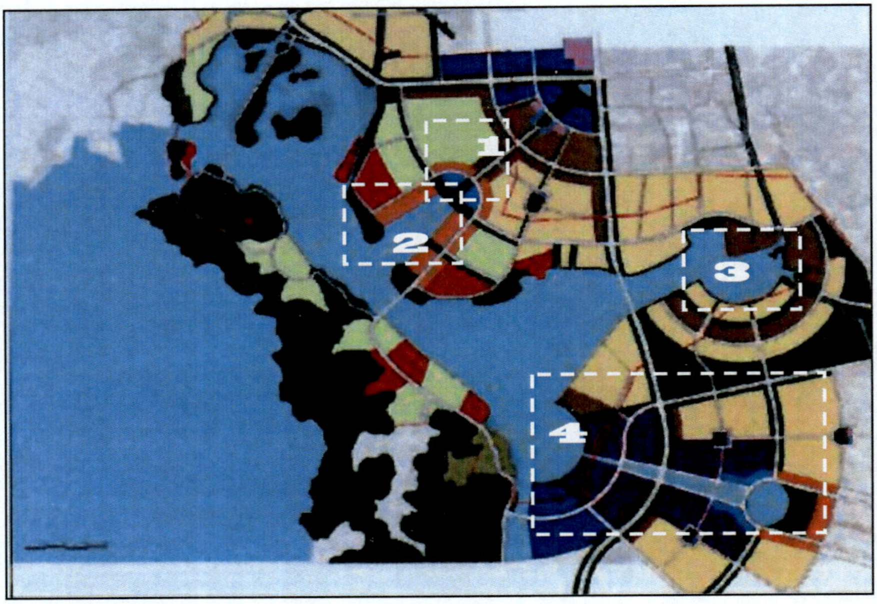

Fig. 8.14 EDAW's conceptual structure plan, option three. *Source* The master plan presentation package of the EDAW team (2002 pre-August: 2), notation by author

Following the provision of the above three options, the EDAW team prepared a package for the 'master plan presentation' (2002 pre-August: 9), which was provided earlier than their completion of the master planning report provided in August 2002. In this package, the master plan suggests a layout as an equal integration of all three options (Fig. 8.15). Such integration enabled a series of circular centres within a principal axis crossing these centres to create an overriding structure for the urban form. The symmetrical deployment of buildings along this axis (Fig. 8.15: 1) and the radial structure of main roads opening towards Li Lake were retained (Fig. 8.15: 2). In addition, fountains were proposed in Li Lake, a significant landscape feature not previously proposed (Fig. 8.15: 3, 4). Moreover, a range of islands were introduced to the central city area (Fig. 8.15: 1), providing a naturalistic water edge and a range of small bodies of water. This contrasts with the third option, in which a large body of water was introduced in the city's central area. It also contrasts with the geometric shape of the axis, circular centre, and symmetrical deployment of buildings. The strategy of islands and a natural-appearing water edge for the central area distinguished the layout of this master plan from the previous.

In the master planning report produced in August 2002, the overall layout had become less geometric (Fig. 8.16). While the characteristic of the city's centre opening to Li Lake was retained, the central axis was nevertheless weakened. Contributing elements, such as the symmetrical deployment of buildings along the principal axis and the circular water features encircled by open spaces, all gave way to a less formal pattern (Fig. 8.16: 1, 2 and 3). Instead of buildings, there was an

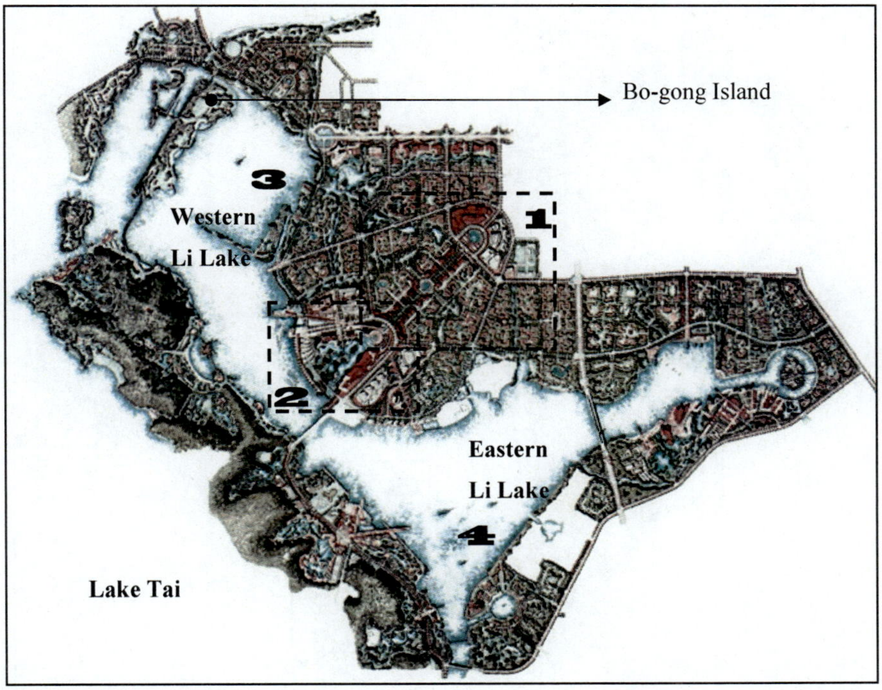

Fig. 8.15 EDAW's initial master plan. *Source* The master plan presentation package of the EDAW team (2002 pre-August: 9), notation by author

increase in open spaces along the axis further introducing landscapes defined by natural elements into the city (Fig. 8.17). A main road added (crossing the centre of the proposed new city) further eroded the original geometric composition as it had survived in the master plan (Fig. 8.15) that was prepared for presentation.

Another significant change to the layout was in the body of water of Li Lake. For Western Li Lake, the original fish pond area (Fig. 8.16: 4), the previously proposed small islands and dams that extended further into the water were removed, doubling the original distance to Bo-gong Island. For Eastern Li Lake (Fig. 8.16: 5), three islands were proposed in the central area, replacing the previously proposed three fountains. This suggests the integration of the planning ideas of the RDUDP team, as these ideas were shown in most plans the RDUDP team produced, including two of the three conceptual structure plans and the final master plan. Changes to other areas also provide further examples of the spatial trend moving away from formal patterns, such as the circular pattern in the eastern portion weakened (Fig. 8.16: 6), as well as the southern (Fig. 8.16: 7).

Figure 8.17 shows a bird's eye perspective drawing of the proposed new city. In this central city area, the proposed islands and the extension of water along the central axis suggest the intention to reference the nearby 'Three Islands' in Lake Tai

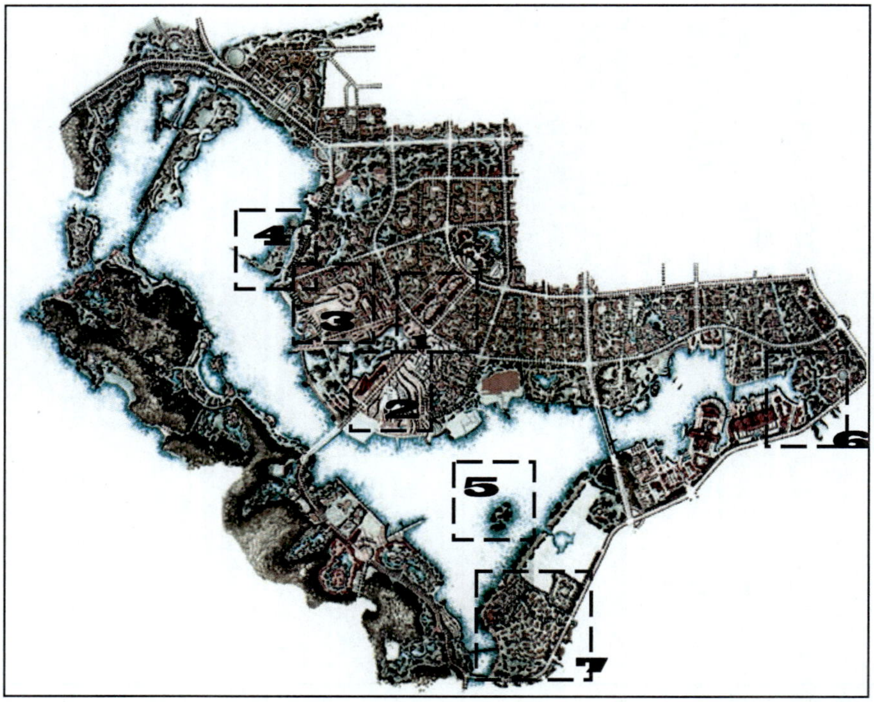

Fig. 8.16 EDAW's final master plan. *Source* The planning report of the EDAW team (2002 August: 22–23), notation by author

and the ancient canals of Wuxi. Also, changes to the layout of the road system in the final master plan suggest an intention to retain existing roads.

Figure 8.18 provides a comparison of the road system before the project started and the system proposed in the final master plan the EDAW team produced in August 2002. In this figure, the thicker lines represent existing roads retained, while the thinner lines indicate new roads proposed in the final master plan. These two road systems suggest that the disappearance of the circular pattern and the appearance of a less geometric layout were for the purpose of maximising the retention of the existing roads.

In addition, the master plan had two key outcomes. The first was the seven 'city-wide planning strategies' developed under the city-wide planning principles. (EDAW 2002 August: V). The seven strategies included a 'tourism strategy' (Fig. 8.19: 1), an 'urban design strategy' (Fig. 8.20), a 'building heights and setbacks strategy' including street character strategy (Fig. 8.19: 2), an 'environmental strategy' (Fig. 8.19: 3), an 'open space strategy' (Fig. 8.19: 4), an 'lake edge strategy' (Fig. 8.19: 5), and a 'transport and access strategy'. According to Winsor (2002), the tourism and environmental strategies were provided by the EDAW San Francisco office in America. This suggests that the other strategies were provided by the EDAW Brisbane office, as well as other professionals based in Brisbane, Australia. Further,

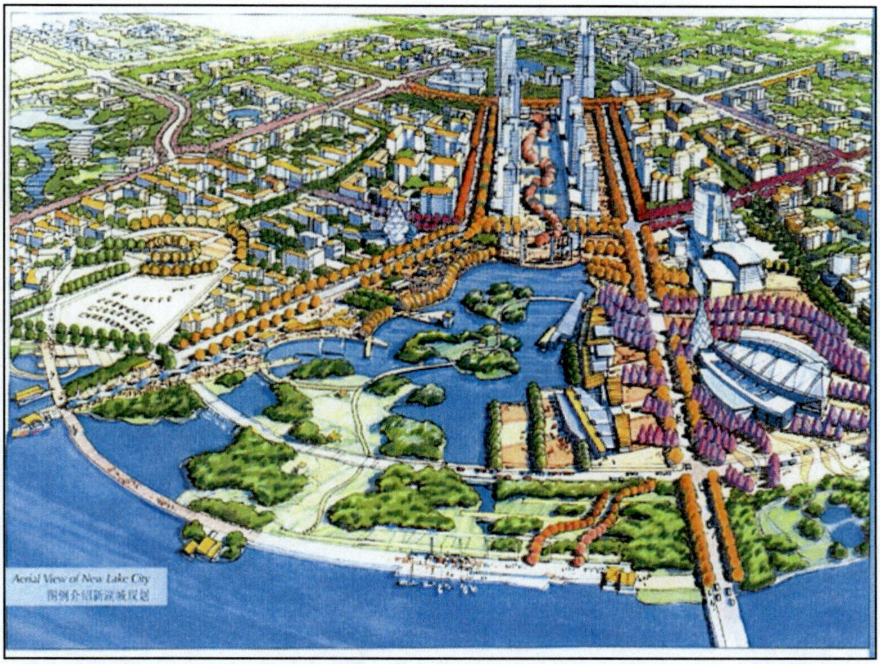

Fig. 8.17 EDAW's final master plan—the central city area. *Source* The planning report of the EDAW team (2002 August: 23)

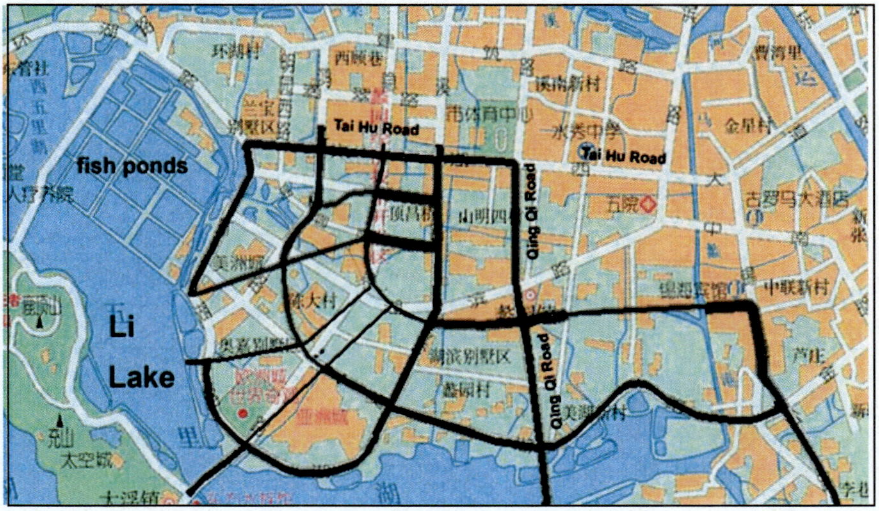

Fig. 8.18 The original road system and EDAW's proposal. *Source* http://www.9654.com/m/wuxi.htm, retrieved 24 May 2010, notation by author, illustrating the road system proposed in EDAW's final master plan

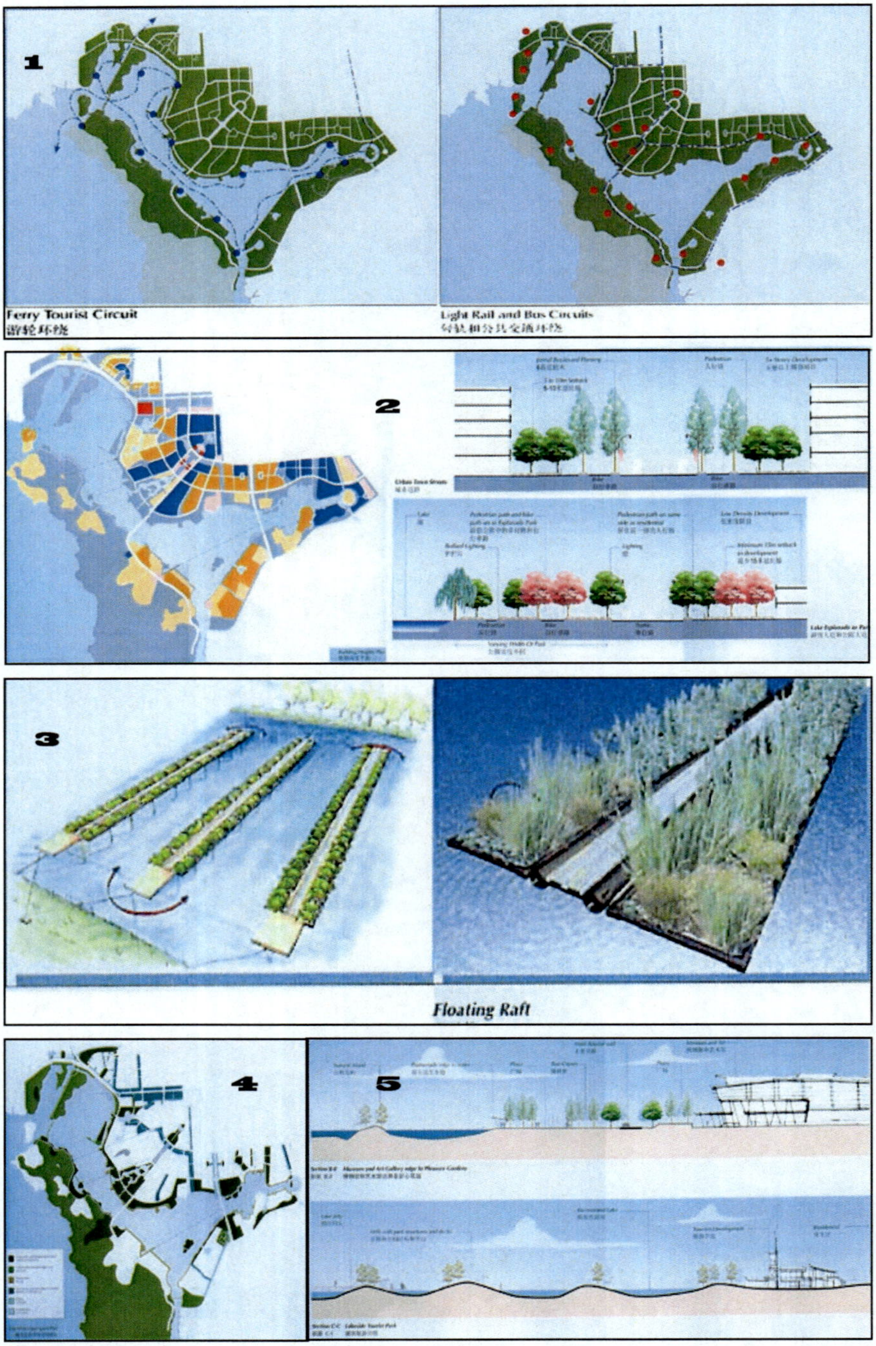

Fig. 8.19 EDAW's city-wide planning strategies. *Source* The planning report of the EDAW team (2002 August: 35), notation by author

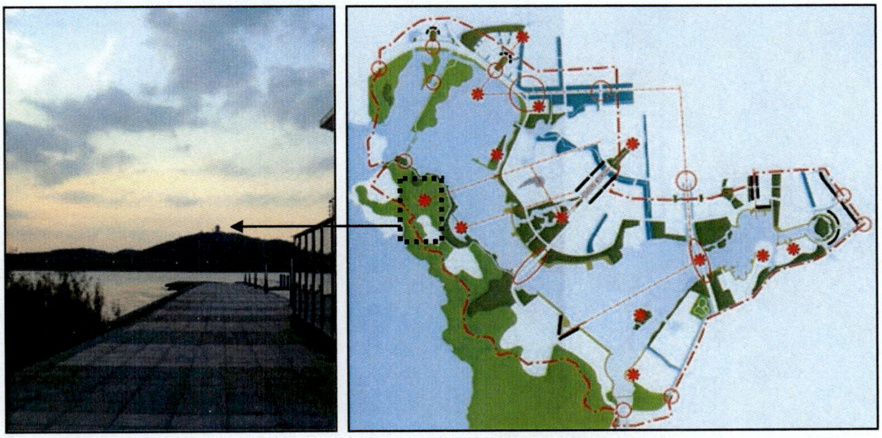

Fig. 8.20 EDAW's urban design strategy for visual connection. *Source* Right—author, photographed in 2007; left—The planning report of the EDAW team (2002 August: 35), notation by author

the urban design strategy suggests the focus of the Australian professionals involved was to create visual corridors connecting to the old landscape located on the opposite shore of Li Lake: the renowned scenic area of Turtle Head Islet. Figure 8.20 shows how the layout of roads in the master plan was proposed to respond to the pagoda located at the top of the hill of Turtle Head Islet, as well as the inlet. Such an idea can even be traced to the second conceptual structure plan option (Fig. 8.13).

The second key outcome was the planning of six precincts of the Li Lake area, including 'Lake Centre' (Fig. 8.21: 1), 'The Quay' (Fig. 8.21: 2), 'Lakeside Village' (Fig. 8.21: 3), 'South Bank' (Fig. 8.21: 4), 'Turtle Head' (Fig. 8.21: 5), and 'the Esplanade' (Fig. 8.21: 6) (EDAW 2002 August: v). Illustrated with plans and perspectives, key places for the six precincts were proposed, and each precinct was proposed, first with planning structure and land use; second, with tourism and recreation activities; third, with landscape character; fourth, with engagement with water; fifth, with living and residential environments; sixth, with massing and density; and lastly, with access and city journey. All were developed from the seven city-wide planning strategies. Compared with the eight precincts proposed by the RDUDP team (Fig. 8.2), a remarkable difference was the developmental vision manifested in each of those precincts proposed by the EDAW team. The eight precincts proposed by the RDUDP team, however, focused more on the identification of the original character of the sites, rather than the development character.

The above city-wide planning strategies and plans of six precincts were explained by texts, extensive plans and sections, which provided a comprehensive and pictorial planning solution to the Li Lake area. From the analysis of the study area to the provision of this master plan, the intention of the EDAW team to integrate local characteristics was mainly embodied in their analysis of the recent evolutional history of Wuxi as a waterfront city, key existing views, and the exploitation of the potential of

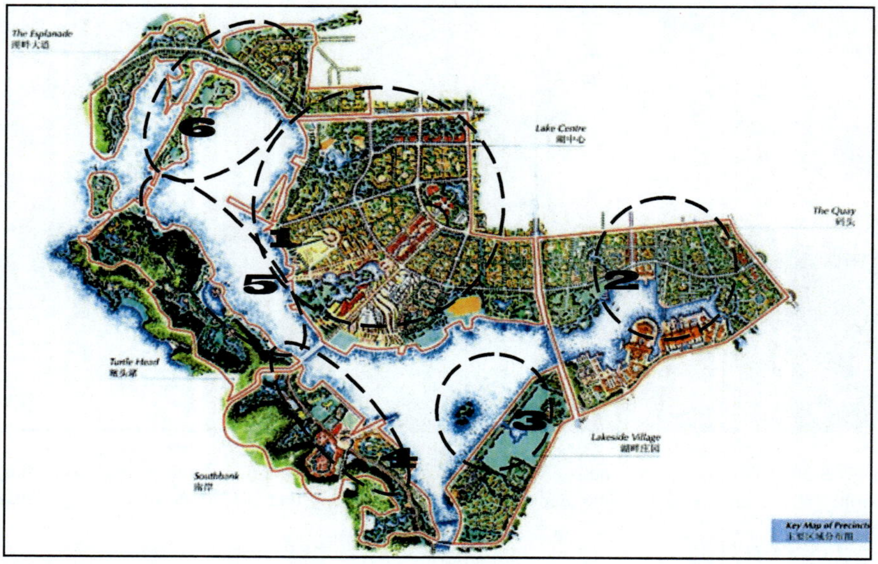

Fig. 8.21 EDAW's proposal for six precincts. *Source* The planning report of the EDAW team (2002 August: 61), notation by author

the landscape characterised by water through a tourism program, ecological planting and treatment of the lake edge. Although their analysis addressed Wu culture and its long history of civilisation, these matters were not further addressed in the planning solutions. All of these suggest that in the planning phase by the EDAW team, local characteristics were emphasised not in considering its past physical forms but in the potential for future development. Elements of the master planning originally provided by the RDUDP team were, however, considered. These included the use of islands in Eastern Li Lake. The landscape design phase that followed was the design of the waterfront for two of the six proposed precincts: 'Lake Centre' and 'the Esplanade' (Fig. 8.21: 1 and 6). These are analysed in what follows.

8.2 The Landscape Design Outcomes

EDAW's master plan had been selected over that of the RDUDP team. According to the official approval issued in December 2002 (Sun et al. 2007: 8), the governmental clients preferred EDAW's master plan for, first, the layout of the road structure that opened toward Li Lake and the scenic area of Turtle Head Islet; second, for the proposal of the six precincts, whose boundaries were relatively independent; and third, the integration of the natural and urban landscape. In the landscape design phase of the northern portion of Li Lake, the team provided a landscape design

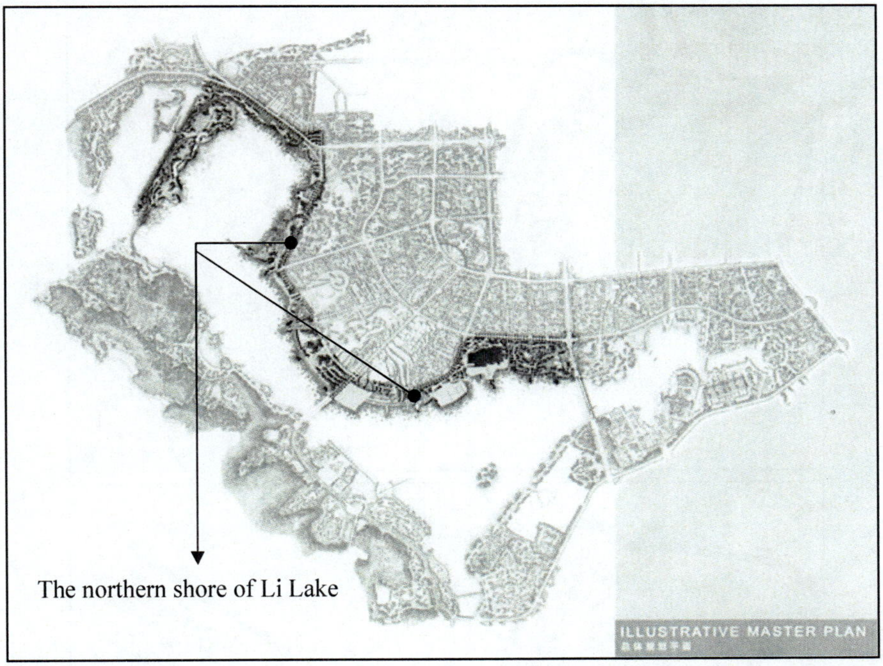

Fig. 8.22 EDAW's working domain for the landscape design phase. *Source* EDAW's landscape design package (2002 December: 2), notation by author

package. Figure 8.22 shows the working area of the landscape design phase of the EDAW team, which covered nearly one-third of the entire Li Lake waterfront.

The landscape design phase shows a range of references to local history. According to the landscape design package, *Five Mile* [Li] *Lake Landscape Design*, dated December 2002, this phase had two stages (Fig. 8.23), the second stage being a development of the first. An overall strategy of the progression from 'natural environment (nature)' to 'new urban environment (city)', and then to 'sustainable environment' was shown in both stages (EDAW 2002 December: 4, 23). Figure 8.23 shows that the landscape designs of nine of the ten components of the waterfront (Fig. 8.23: 1–9) were different between the two stages. Yet, their references to local history were similar. Ways to refer to local history in these designs are analysed further in the following sections.

References to local history in the landscape design were drawn from four elements. The first was the mythical story of Li Lake, in which a wakening dragon created the lake. This story was interpreted through using Dragon as the name for the proposed island, using sculptural art pieces and paving inlays that tell stories, and imitating the physical form of a dragon in design (Fig. 8.24). The second was the story of the two lovers, Fan Li and Xi Shi, from which Li Lake was named. Ways to interpret this story in design were similar to those used for the first. Figure 8.24 shows that these two local stories were interpreted in the landscape design of an island proposed to

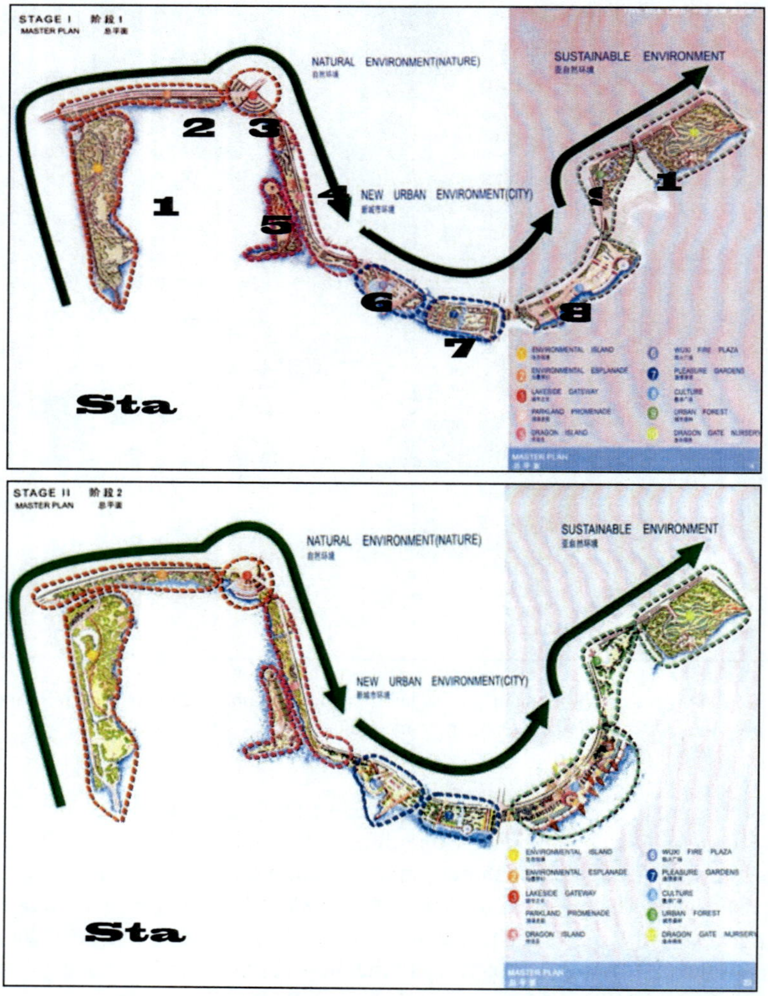

Fig. 8.23 EDAW's two stages of landscape design. *Source* EDAW's landscape design package (2002 December: 4, 23), notation by author

replace the original fish pond area. The island was named Dragon Island in Stage One. Sculptures of dragons were indicated in this design. In Stage Two, the island was renamed Legend Island, where the planting design was changed to a geometric layout.

Comparisons with the island proposed in the planning phase show that the response to these local stories started after the planning phase. Prior to this, the island had been entitled Maze Island and was composed of a Pyramid Maze, Flora Maze, and Sculpture Garden and Maze (Fig. 8.25). The onset of the integration of local elements in the design can be related to the clients' hosting of the Cultural

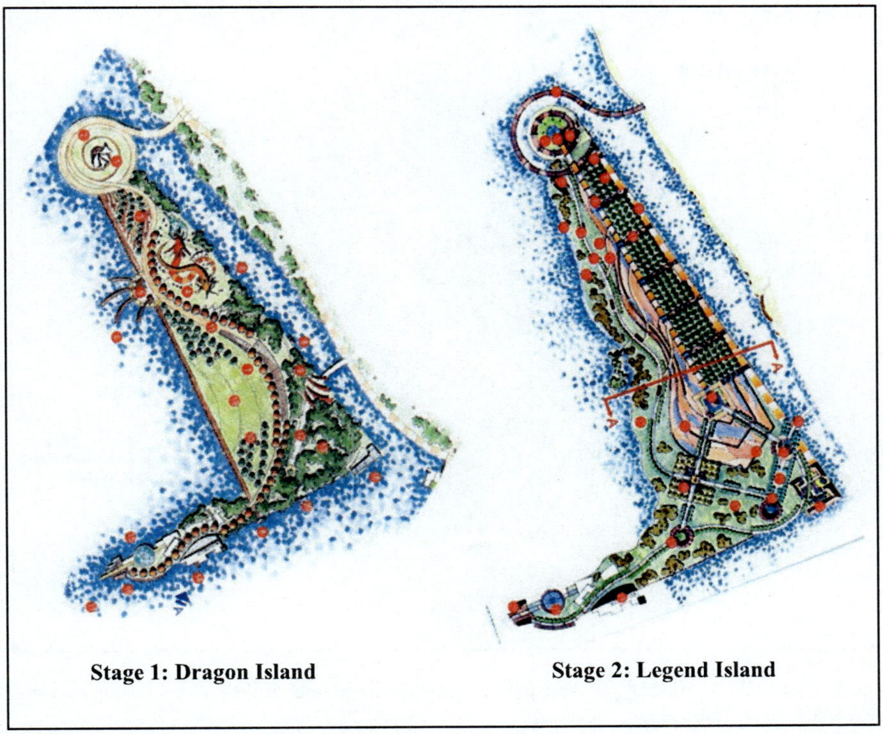

Fig. 8.24 EDAW's design of an island—the local legend. *Source* The landscape design package of the EDAW team (2002 December: 11, 30)

Symposium held on 20 September 2002 in Wuxi (EBOCWJ 2002), which followed the production of the master plan report in August and preceded the production of the landscape design package in December. The news issued by EBOCWJ (2002) also shows that in the Cultural Symposium, artists and poets were gathered to discuss the names of twenty-four scenic spots around Li Lake. This suggests that the naming strategy of the EDAW team was not used particularly in the planning phase, but it was later influenced by the ideas of local experts that were offered in the Cultural Symposium. This news, however, did not address the presence of the EDAW team in the symposium. This suggests that the team was not directly engaged in this first symposium for the cultural construction of the Li Lake area.

The third local element was the ancient mining history of Wuxi, for which the city was named. Ways of interpreting this story included recalling the ancient scenario of smelting and founding tin by creating specific functional areas in the design, such as fire plazas (Fig. 8.26). This suggests the intention to incorporate the mercantile culture that distinguishes the local culture as Wu culture.

The fourth local element was the agricultural and aqua-cultural aspects of Wuxi, which were interpreted through the symbolic use of the patterns of rice paddies, waves and ships. Figure 8.27 shows the layout of the planting design for the 'orna-

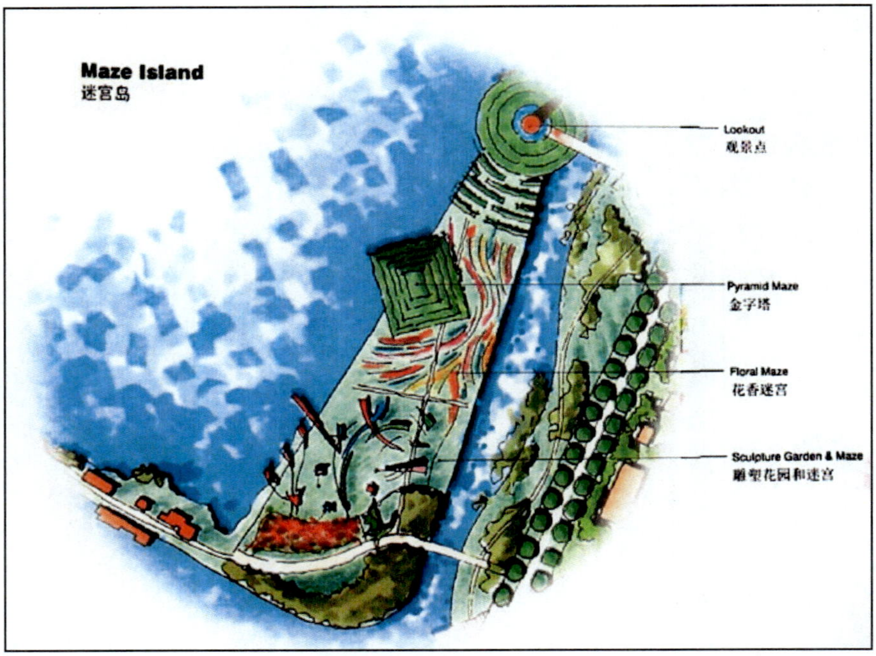

Fig. 8.25 EDAW's proposal for the island in the planning phase. *Source* The planning report of the EDAW team (2002 August: 70)

Fig. 8.26 EDAW's design of waterfront plazas—local history. *Source* The landscape design package of the EDAW team (2002 December: 33)

mental nursery', suggesting lake waves; and the trapezoidal form of the 'Dragon gate sculptures', alluding to boats returning to a harbour.

Of the above four local elements that inspired the landscape design of the northern shore of the Li Lake, three are legends or folklore embedded in the local ancient history of Wuxi, one of which is concerned generally with the local agrarian peasantry and fishery. The design package shows that the former three local elements influenced

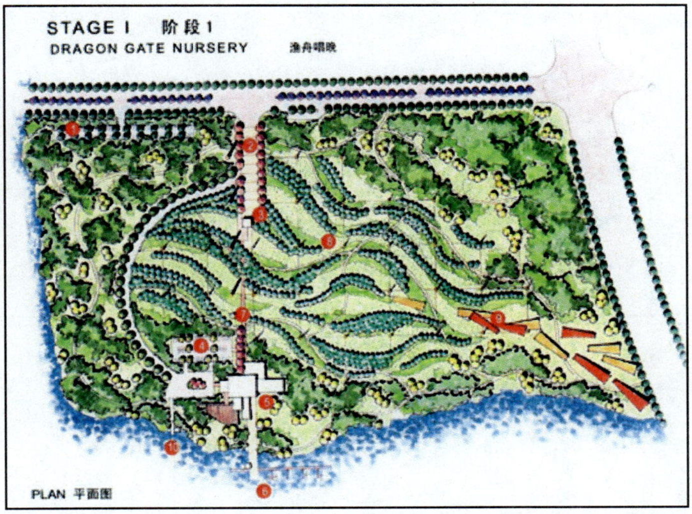

Fig. 8.27 EDAW's design of a waterfront park—local fishery culture. *Source* The landscape design package of the EDAW team (2002 December: 20)

the design of some key spaces in the waterfront area, such as Dragon Island and Fire Plaza. These local references also suggest the emphasis in design on interpreting a heroic part of local history.

In addition, the landscape design also proposed a series of plazas, such as the Market Plaza and Blossom Plaza to accommodate family and community life. According to the landscape design package, community life such as practicing tai chi, dancing and morning exercises (EDAW 2002 December: 14, 33) were addressed, expressing concern for the typical Chinese morning lifestyle that has lasted for decades and centuries. These activities however, are common not only in Wuxi and the Jiang-nan region, but also in all major Chinese cities from Beijing in the north to Hong Kong in the south. This suggests a general reference to Chinese people, rather than to a specific local community of the Li Lake area.

In the landscape design phase, the interpretation of native flora and fauna, however, is not apparent, except for the proposal of transforming the existing Bo-gong Island into an 'environmental island' for education and learning, which suggests the use of native species there. The design also proposed an Urban Forest, Dragon Gate Nursery, and Parkland Promenade. However, ways in which local plants would be used were not specified. The concept of the planting design can only be traced from design drawings through the forms and colours illustrating plants. Of the illustrated forms, some suggested local species such as the camphor tree and metasequoia; some, however, suggested tropical species (Fig. 8.28). Moreover, a variety of colours were used to express the idea of using plants. For instance, trees were often coloured in bright purple, suggesting the idea of using a single tree species to provide a contrast or rhythm. However, the landscape design package did not contain a plant palette.

Fig. 8.28 EDAW's planting design in the landscape design phase. *Source* EDAW's landscape design package (2002 December: 6, 11, 17, 34)

Therefore, the use of colours became a dominant theme, signifying the natural characteristics of the area in the future. In the mild and evergreen Jiang-nan region, however, tree species are mainly identified by forms rather than by colours. Moreover, using plants for their associated moral, picturesque and poetic meanings is a traditional way of making gardens and landscapes in China. This important part was not addressed in this phase. Instead, the message expressed through design drawings suggests the focus on the plants' horticultural value, and the reference to the Western tradition of topiary (Fig. 8.28: bottom).

8.3 The Landscape Design Development Outcomes

The working domain of the landscape design development was the Western Li Lake waterfront, including three areas. Documents for the landscape design development, in fact, were created at a depth close to what is normally considered construction documentation, including landscape materials, finish schedules and lists of plant species. This explains how the landscape design development of the Western Li Lake waterfront paralleled its construction. Figure 8.29 shows the three areas of the landscape design development phase, including first, Wuxi Lakefront part one, completed in the first half of 2003; second, Wuxi Lakefront part two, completed in July 2003; and third, Fishing Dragon Island, completed in November 2003.

A reading of the landscape design development packages, *Wuxi Lakefront, stage two*, dated July 2003, and *Fishing Dragon Island*, dated November 2003, reveals

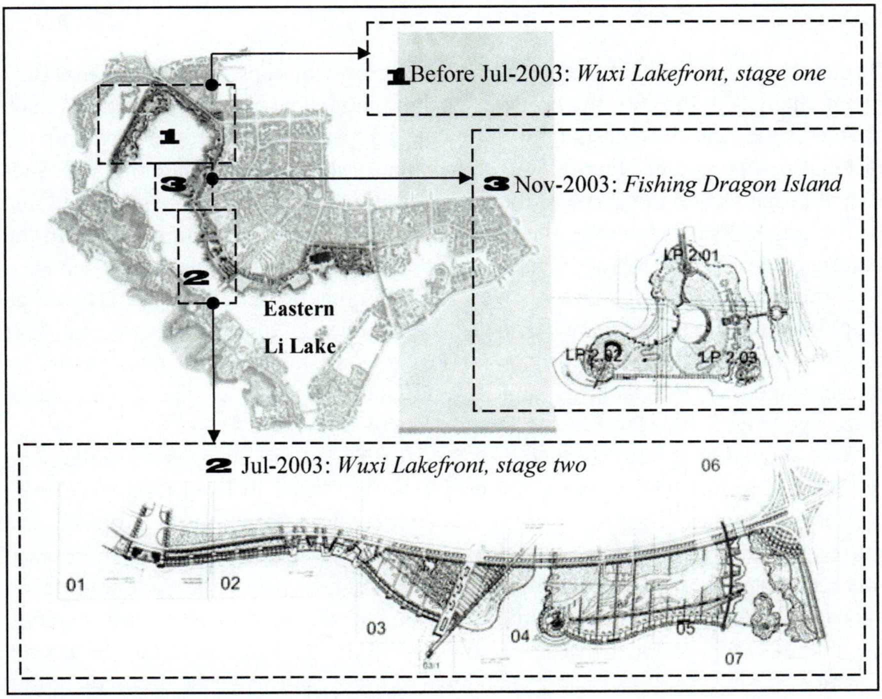

Fig. 8.29 EDAW's working domain for landscape design development. *Source* The landscape design package and design development packages of the EDAW team (2002 December: 2; 2003 July: cover sheet; 2003 November: levels and guiding plan), notation by author

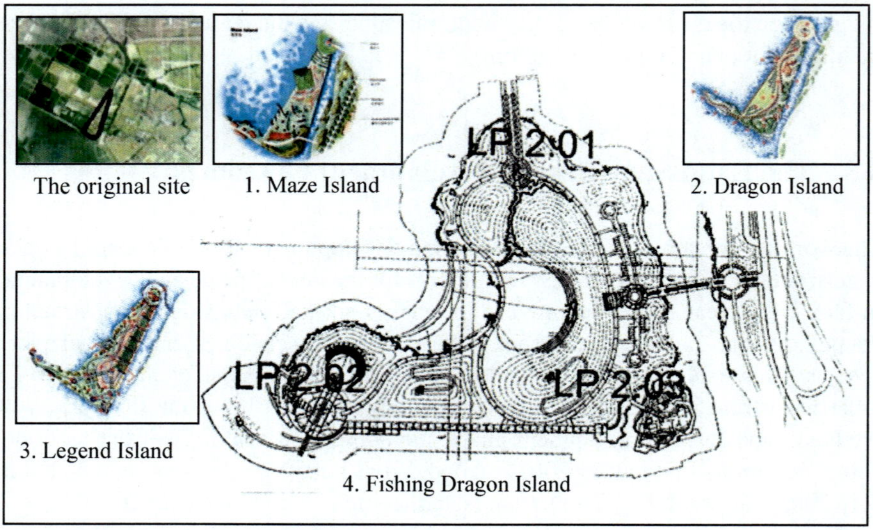

Fig. 8.30 EDAW's designs of the island over phases. *Source* EDAW's planning report, landscape design package and landscape design development packages (2002 August: 13, 70; 2002 December: 11, 30; 2003 November: levels and guiding plan), notation by author

further integration of local characteristics into the previous design, resulting in consequent changes to those of the previous design phase. The clearest example of such change is the landscape design development of an island proposed for replacing the original fish pond area. Figure 8.30 shows that in the previous planning and landscape design phases the shape of the proposed island continued to be similar to that of the original site. However, the retained shape was changed dramatically in the landscape design development phase, with similar dramatic changes to its elements. Two prominent landscape features were added in this phase: a sandy beach and an amphitheatre. The name of the island also shifted through four iterations: from Maze Island of the planning phase (Fig. 8.30: 1); Dragon Island of the landscape design phase, stage one (Fig. 8.30: 2); and Legend Island of the landscape design phase, stage two (Fig. 8.30: 3) to Fishing Dragon Island (Fig. 8.30: 4).

The above changes to the name of the island with the changes from planning to design development, suggest that design ideas shifted to local references from western references. The use of western references was shown in the proposal of a maze and sculpture garden as planning proceeded to design. In addition, references to local characteristics were also shown in the design of hard landscape of the waterfront, such as shelters, sculptures, buildings, pavement and outdoor furniture. Overall, the intention to use local fishery culture as the design inspiration can be traced. Figure 8.31 shows ways of using such local references of fishery culture, including the physical forms of fish and waves, and fishing instruments such as fishing-net and fishing-rod.

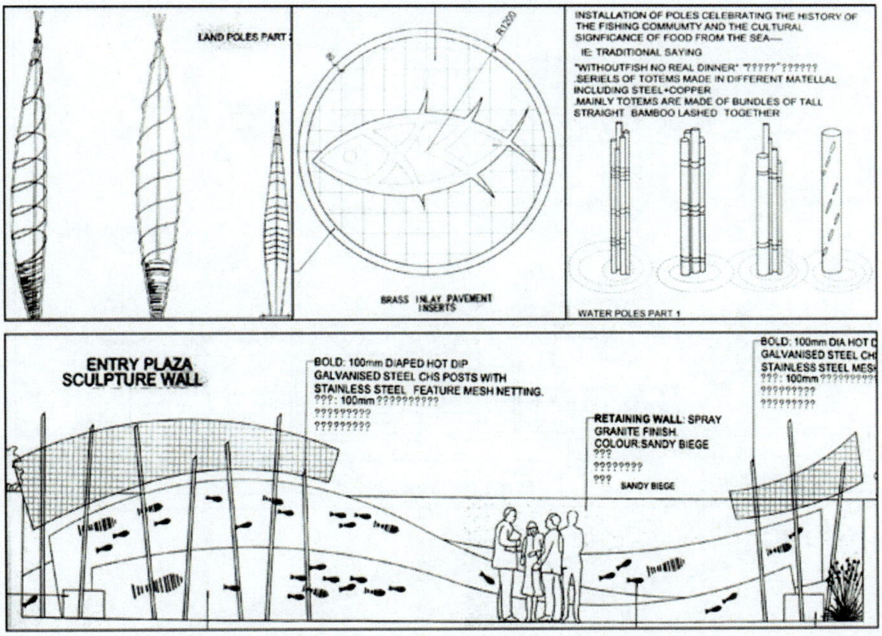

Fig. 8.31 EDAW's design of hard landscape—local fishery culture. *Source* The landscape design development packages of the EDAW team (2003 July)

Such references to local fishery culture were also embodied in architectural designs, such as the design of a private yacht club, a maritime museum and coffee houses. In all, an intention to use sails and waves as local references can be read (Fig. 8.32). Of the various small waterfront buildings, as well as shelters, the composition of timber frames and metal structures suggest a character similar to buildings of the Darling Harbour, Kings Wharf, which are renowned waterfronts of Australia. While the above architectural and art forms suggest the use of physical forms to interpret local characteristics, the adopted modern materials and compositional methods of tea house, cafes and fishing shelters also suggest Australian elements.

The phase of landscape design development also suggests the consideration of local species. Changes to the idea of planting design of the previous design phase were shown in the layout of planted areas. Compared with plans provided in the landscape design phase, changes to the layout of planted areas were accompanied by dramatic changes to the layout of plazas and paths (Fig. 8.33). Comparisons between the design and design development phases show the decrease in organic patterns in layout of the planted areas, plazas and paths, such as the disappearance of meanders along mounds and across the parklands. In the design development phase, the parallel straight paths enhanced geometric order (Fig. 8.33: bottom). These paths cut through the planted areas and present a composition less organic than that expressed in the landscape design phase (Fig. 8.33: top).

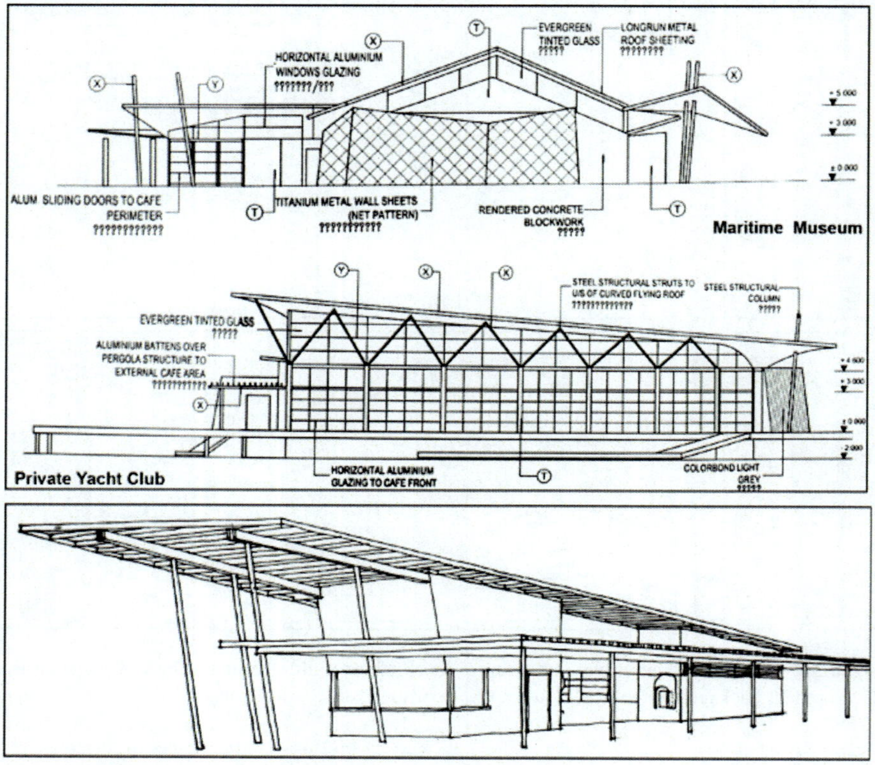

Fig. 8.32 EDAW's architectural design—local fishery culture. *Source* EDAW's landscape design development packages (2003 July)

In addition, local plant species were specified. In the plant lists of the design development packages (EDAW 2003 July; EDAW 2003 November), most of the selected trees and shrubs are common species and popular for landscape use in the Jiang-nan region, such as *Cinnamomum camphora*, *Liriodendron chinensi*, *Ginkgo biloba*, and *Canna generalis*, *Nerium indicum*. The plant lists also included a range of native waterfront species. However, tropical species were also selected for Fishing Dragon Island, such as *Phoenix canariensis*, *Washingtonia filifera*, and *Phragmites australis*. Except for the abovementioned tropical species, the selected plants were Jiang-nan region-based or had been localised and popular in landscape use in that region for a long time. This contrasts with the illustrative images of planting design in the previous phase. This suggests the more specific identifications of local natural characteristics in the design development phase, but with exotic plants incorporated as accent planting to provide contrast. This also suggests the EDAW team's acquirement of knowledge of local species started in the design development phase.

Along with the interpretation in design of a heroic part of local history and common natural characteristics, the design development phase paralleled the construction process of the Western Li Lake waterfront, which started in early 2003. In this phase,

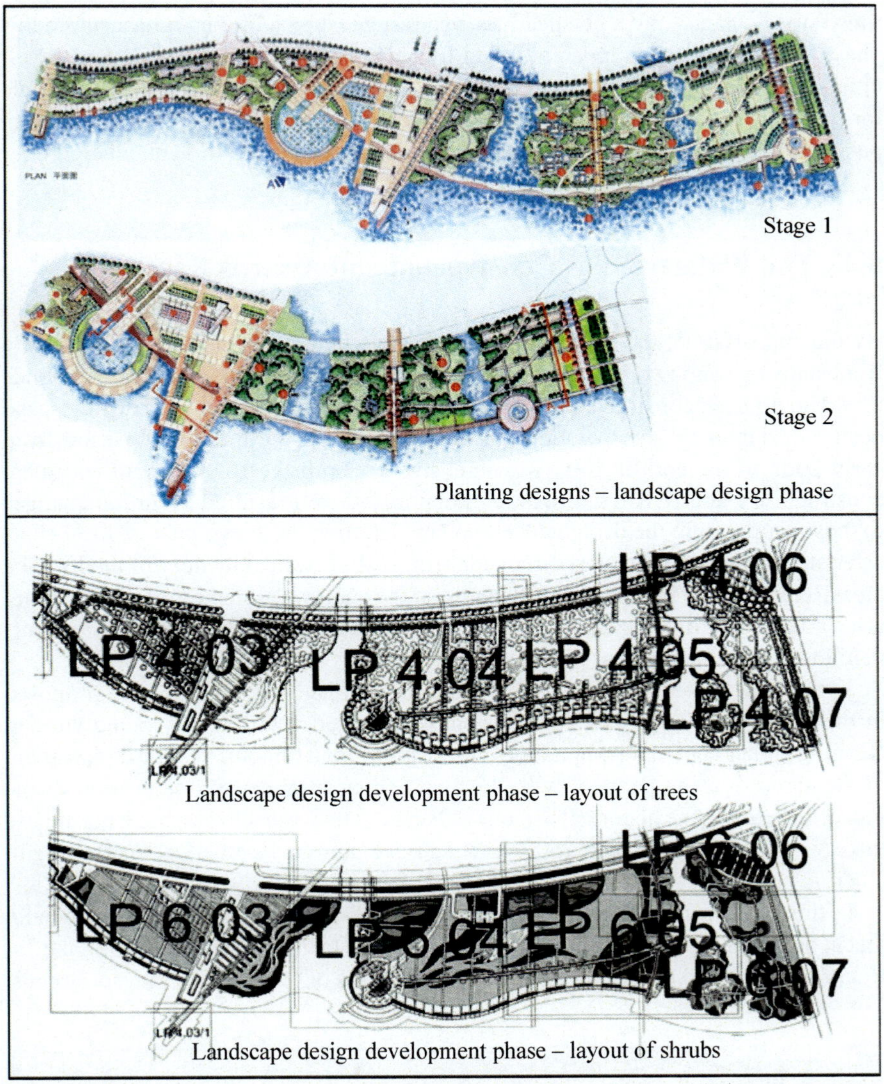

Fig. 8.33 EDAW's planting designs—changes over phases. *Source* The landscape design and design development packages of the EDAW team (2002 December; 2003 July)

between the production of two design development packages in July and November 2003 respectively, part of the Western Li Lake waterfront parklands was constructed and opened to the Wuxi public in October 2003. This suggests that the design development phase was an intensive and overlapping working phase for the EDAW team. Substantial changes can be read in the previously discussed layouts of the proposed island, the planted area, plazas and paths. Contrasting with the most computer-aided drawings in the design development package, some architectural designs were simply illustrated by hand-drawn perspectives, again suggesting a limited time span. In this phase, the EDAW team apparently had four major tasks: first, revising and developing the previous landscape design; second, developing the design to a level suitable for construction; third, generating design concepts for architectural forms, including public arts; and lastly, construction supervision, which started in September 2003.

8.4 The Waterfront as Constructed and Awards Received

By the end of 2005, the Western Li Lake waterfront, including the proposed lake front stage one and two, and the island proposed in the previous polluted fish ponds area, had all been constructed as parklands and opened to the public. The landscape design development and parallel construction suggest the three-year period from early 2003 to the end of 2005 was intensive for both the EDAW team and local construction teams. Also, the construction process itself resulted in further changes to those proposed in the design development. Through these two phases, local characteristics were further materialised, confirming the physical identity of the Li Lake area. Referred to by both the EDAW team and the clients, the shelters of an amphitheatre and of a boardwalk along the lake front became the main landscape statements defining the new identity of the area.

In the constructed waterfront, the name of the Fishing Dragon Island proposed in the landscape design development phase, changed again, finally being entitled 'Fishing Father Islet', and now adopted in the guide map for tourists. The replacement of the name 'Fishing Dragon' with 'Fishing Father' after construction suggests the local emphasis on the historical personage Fan Li, who wrote the first book in Chinese history on fish farming about 2500 years ago. It again shows an effort to referring to a heroic part of local history.

In this completed Fishing Father Islet (Fig. 8.34), the sandy beach is artificially made, but attracts people to playing with the water. On the beach, a range of Lake Tai Rocks are scattered (Fig. 8.34), suggesting elements of traditional Chinese gardens.

Fig. 8.34 The constructed Fishing Father Islet. *Source Unveiling the beauty of Li Lake: Wuxi well on its way towards ecological well-being* (WMCPC 2007: 15), notation by author

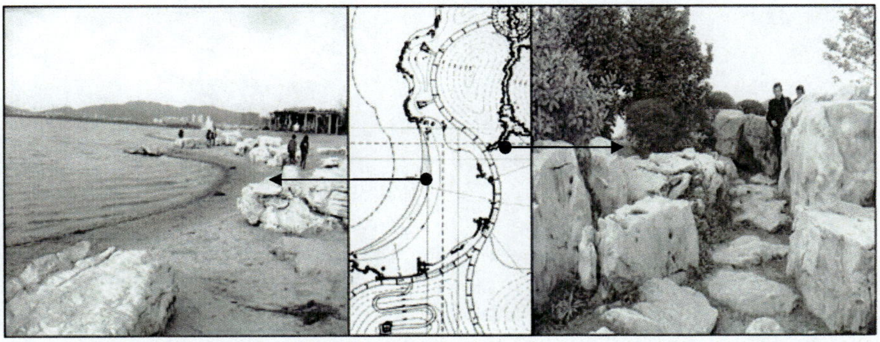

Fig. 8.35 The constructed waterfront—an artificial sandy beach. *Source* Author, photographed in April and November 2007; and, EDAW's landscape design development packages (2003 November)

Also, bridges of Li Dyke show a resemblance to those depicted in ancient Chinese landscape paintings, as was discussed in Chapter Six. These bridges connect the islet to Bo-gong Island. These physical entities form an expression that can be traced to the characteristics of Wuxi and Jiang-nan, and are arguably attempts to respond to a perceived sense of place.

Compared with the island proposed in the design development phase, the idea of a sandy beach was enlarged in the constructed Fishing Father Islet (Fig. 8.35). The sandy area was extended substantially. Such a change provided a more open and less enclosed spatial outcome to Western Li Lake. Also, more rocks than those proposed in landscape design development were deployed in the central area of the beach. According to the construction supervisor of the EDAW team (Brookbank 2007: personal communication), these rocks were a solution to the over-supply of rocks from other parts of the project sites. As an influential element of traditional Chinese gardens, a regional sense of place of Jiang-nan is expressed by these Lake Tai rocks. These rocks enrich the texture of the beach and provide sitting areas.

Ways to deploy these Lake Tai rocks in the waterfront differed from those used for traditional Chinese gardens, such as those in the nearby Li Garden (Fig. 8.36: top) and those signifying Taoist meanings. In addition, these rocks are also composed with trimmed shrubs of *Euonymus japonica* and the colourful *Acer palmatum* and *Rhododendron simsii*. These landscape features define a semi-open place within the open waterfront area (Fig. 8.36: middle and bottom) and suggest the style of Japanese gardens, rather than those traditional of Wuxi and Jiang-nan.

Differing from the landscape design development, the constructed Fishing Father Islet had added to it a 1200-metre-long dyke: Li Dyke, connecting the islet with Bo-gong Island (Fig. 8.37). The construction of the dyke was completed in the end of 2005 and seen as a milestone of the completion of water quality improvement of Li Lake and of waterfront renovation of Western Li Lake. The design of the dyke however, had not been included in any of the previous phases of EDAW's work from planning to landscape design and to landscape design development. According to

Fig. 8.36 The constructed waterfront—ways to use Lake Tai Rocks. *Source* author, photographed in April 2007; and, *Unveiling the beauty of Li Lake: Wuxi well on its way towards ecological well being* (WMCPC 2007: 15)

the article produced by Wuxi Landscape Design Institute (Sun et al. 2007: 110), its design was directed by Sun, the head of WMCPC.

Although not included in the previous phases of EDAW's work, the intention to extend Fishing Father Islet closer to Bo-gong Island is shown clearly in their initial master plan (Fig. 8.38). The creation of Li Dyke also suggests the practical

Fig. 8.37 The constructed Li Dyke—the traditional form of bridge. *Source Unveiling the beauty of Li Lake: Wuxi well on its way towards ecological well being* (WMCPC 2007: 24)

Fig. 8.38 The constructed waterfront—local history in architectural forms. *Source* Author, photographed in November 2007

collaboration between the EDAW team and the WMCPC in their parallel design development and construction phases. The creation of the dyke provided a new access to the other side of Li Lake, the famous Turtle Head Islet, and the longest trip possible to cross the lake. Furthermore, the dyke comprises arch bridges in traditional style, expressing a visually recognisable sense of place. In addition to the added Li Dyke, the Maritime Museum that had been documented was not built and, instead, an old building with Chinese traditional style roof was retained and renovated as the tourist centre (Fig. 8.38: 1). New pavilions containing memorial steles were built there with traditional architectural forms (Fig. 8.38: 2).

Changes to EDAW's landscape design development were also shown in functional areas proposed for different types of activities. Between the sandy beach and the planted area of an open lawn and leafy mounds, an amphitheatre had been proposed, sheltered with tensioned membrane structures. However, in the constructed waterfronts, the amphitheatre and the shelter were not built. Instead, facilities for morning exercise and children's play, which had not been included, were now deployed in that area. This alters a supposedly relatively quiet area to an active place brightened by the colourful children's play area (Fig. 8.39: 1). The previously proposed amphitheatre

Fig. 8.39 The constructed waterfront—changes to an amphitheatre. *Source* author, photographed in 2007; *Unveiling the beauty of Li Lake: Wuxi well on its way towards ecological well being* (WMCPC 2007: 15); and, the landscape design development packages of the EDAW team (2003 November)

Fig. 8.40 The constructed waterfront—local culture in architectural forms. *Source* Author, photographed in April 2007; and, *Unveiling the beauty of Li Lake: Wuxi well on its way towards ecological well being* (WMCPC 2007: 15)

and its shelter were relocated in the other part of the Western Li Lake waterfront (Fig. 8.39: 2).

A few hundred metres away from the Fishing Father Islet a shelter is constructed in the form and composition similar to that proposed in the design development phase for the island (Fig. 8.40: 2). Such a form, according to the EDAW team (Brookbank 2007: personal communication), was inspired by the flying birds of Lake Tai. The shelter is located along the water surface of Li Lake. It forms a close place for watching a 120-metre-high fountain in Western Li Lake (Fig. 8.40: 3), which was designed by fountain specialists in America. The architectural form of the shelter becomes one big landscape statement.

Another landscape statement is the shelter for a boardwalk on the water surface of Western Li Lake (Fig. 8.40: 1). According to the EDAW team (Brookbank 2007: personal communication), the architectural form of the shelter was inspired by traditional fishing boats on Lake Tai. Throughout this research, it becomes apparent that this architectural form was designed by the Brisbane architect Szczepan Urbanowski (Harvison 2007: personal communication). This further supports the identified sense of Australia that is expressed through buildings of the Li Lake waterfront. Overall, along the Western Li Lake waterfront, three visually outstanding landscape features are established (Fig. 8.40: 1, 2 and 3), including first, the sheltered boardwalk imitating sailing boats; second, the sheltered amphitheatre imitating flying birds; and third, the fountain at a height of about 120 metres. The idea of deploying a fountain in the lake was also shown in EDAW's previous planning and design work.

In addition to the three monumental landscape statements, local characteristics are also expressed in artworks. Bronze statues of historical personage and artworks, which were not shown in EDAW's design, are installed in the area around Li Dyke (Fig. 8.41: top) interpreting local fishery culture and relating to the legend of the

Fig. 8.41 The constructed waterfront—local history and fishery culture in art forms. *Source* Author, photographed in April and November 2007

hero, Fan Li, who is renowned in Chinese history. Figure 8.41 also shows how pavements and lighting facilities reflect the local fishery culture. Such interpretation of local specificities, as analysed previously, started in the phase of landscape design development.

Another important dimension in the expression of local characteristics is how people actually use the waterfront. Of the long-established local outdoor activities, some are specific to the Jiang-nan region, especially Wuxi, such as water-associated activities. These activities are encouraged by the artificial sandy beach. Although prohibited for swimming by local authorities at that time because of the water quality, the sandy beach stimulated the willingness to engage with water (Fig. 8.42).

In contrast, many of the expansive paved plazas are empty of use, with the exception of holidays, weekends and tourist seasons (Fig. 8.43). A retailer who sells food

Fig. 8.42 The constructed waterfront—local activities in the sandy beach. *Source* Author, photographed in April 2007

at the waterfronts said that most park users are residents rather than tourists, and the area is popular only during the summer and weekends. Although coffee bars, water-sports clubs and restaurants are scattered throughout the plazas, the outdoor public spaces with hard surfaces remain un-activated. This also suggests local preferences for the less built and constructed areas, or areas artificially made with a naturalistic appearance, such as the manmade sandy beach. Overall, places with a naturalistic appearance are more popular than the highly built, paved and facilitated areas in the Li Lake waterfront parklands.

The constructed waterfront parklands of Western Li Lake suggest that some of the significant ideas that had not emerged in the planning, landscape design, and landscape design development stages emerged in the construction stage, such as the topography of Fishing Father Islet, the Li Dyke, and statues and art forms reflecting historical personages and their legends. It could be argued that these changes were proposed by local stakeholders, such as the government clients (as shown in the Li Dyke), local experts (as shown in naming and the statues), and local design and construction teams. The latter includes local ways of construction and the availability

Fig. 8.43 The constructed waterfront—local activities in paved plazas. *Source* Author, photographed in April 2007

of local products. For example, the Australian landscape architects from the EDAW team, who acted as construction supervisors (Brookbank 2007: personal communication), as well as the principal of EDAW Brisbane office (Harvison 2007: personal communication), all explained their concerns with local conventions of construction, such as the use of manual rather than machine operations for carrying and cutting paving materials, the limited time for construction, and relying mostly on on-site experience than on design documents. They believe that these all affected the proposed outcomes to varying degrees.

By contrast, local contractors (WUSRS & LTO 2007: 109–133) believed they had made sufficient effort to comply with design documents to guarantee construction quality. Local contractors concluded they had three major issues to overcome (2007: 109–133). First, the naturalistic patterns in design drawings had to accommodate specific site conditions and the available construction tools, making on-site modification inevitable. Second, new items emerged during the construction process without prepared design or construction drawings. Third, the limited time allowed

Table 8.1 The constructed waterfront—awards received. *Source* The websites of AILA, TWC, ULI, retrieved 2005 and 2006; and, *Urban Science Research of Wuxi: Li Lake Special* (2007: bottom page)

Name of awards		When	Where and by Whom
1	2005 Excellence on the Waterfront Awards	11/11/2005	The Waterfront Centre, America
2	2005 Design Awards in Landscape Architecture	2005	AILA Queensland state, Australia
3	2005 Excellence of Municipal Sculpture Design in China	11/05/2006	China
4	2006 ULI Asia-pacific Awards of Excellence	28/07/2006	Urban Land Institute, Tokyo, Japan
5	2006 Awards of Model City of Water Environment Renovation	08/2006	The Ministry of Construction of China

for construction required negotiations between local contractors to carry out work but avoid clashes on-site.

Although the abovementioned limitations are embodied in the construction process and manifest themselves in the changes to EDAW's planning and design, since late 2005, the renovated waterfront has received a number of international and Chinese national awards. Table 8.1 lists the awards that are relevant to the services provided by the EDAW team, the Australian landscape architects, and other Brisbane-based professionals. Included in the three international awards, the shelter of the boardwalk was cited as the key evidence of incorporating local characteristics into the landscape design. This architectural feature also won the national award in China for being an excellent design of city sculpture. This further suggests that the interpretations of local characteristics are seen as the key success in international and Chinese national awards systems, where visual forms play a key role in such interpretation. This is further discussed in Chapter Nine.

8.5 Conclusion

During the four-year period from the beginning of 2002 to the end of 2005, the participation of the two international design teams in the development of the Li Lake area in Wuxi, China comprised planning, landscape design and landscape design development and construction supervision. Along with this process, the identification of local characteristics, as shown in the principal planning and design documents of the two international design teams, suggests an inconsistent trajectory.

In the planning phase, there were clear intentions of the RDUPD team to incorporate local elements, especially those outstanding in the local history of Wuxi, including the surviving urban form, and the traditional architectural forms of Chinese gardens, and Jiang-nan water villages and towns. The team also emphasised consideration of local residents as well as tourists in functional planning. In contrast, such intentions were not clearly shown in the planning outcomes of the EDAW team. Instead, the concern with local characteristics was fundamentally centred on the city's evolution as a waterfront city, and the potential of such a character for improving the local tourism economy. Therefore, in terms of interpreting the characteristics of Wuxi and Jiang-nan, the RDUDP team looked towards the past of Wuxi and the region to draw references from, while the EDAW team looked towards the future and modernisation to draw references from, and found little to do with the ancient and traditional forms.

In contrast to the clients' choice for the master plan provided by the EDAW team, in the later planning phase and landscape design phases of the EDAW team, there were the beginnings of integrating outstanding elements of local history into planning and landscape design. The clients' suggestion of integrating the merits of the master plan produced by the RDUDP was also followed. The great sandy beach, boardwalk, three islands, as was proposed by the RDUDP team in its planning phase, were all realised in Li Lake and its waterfront. The constructed Western Li Lake waterfront has also retained old buildings in Chinese traditional architectural forms, with statues and artworks to recall heroes in local history. Important landscape areas were also named and renamed to recall legends of historical personages and local folklore.

References

Brookbank, P. (2007). Interview with Yun Zhang on 06 November at EDAW Shanghai office.

Editorial Board of *China Wuxi* Journal. (2002). Brief introduction of detailed planning of Li Lake New City. Retrieved May 7, 2008, from http://www.wuxi.gov.cn/intowx/survey/admin/20069/20069527768.shtml.

EDAW. (2002_pre-August). New Lake City, Wuxi China, Master Plan Presentation.

EDAW. (2002_August). Wuxi New Lake City, Master Plan Report.

EDAW. (2002_December). Five Mile Lake Landscape Design.

EDAW. (2003_July). Landscape Design Development Package: Wuxi Lakefront.

EDAW. (2003_November). Landscape Design Development Package: Fishing Dragon Island.

Harvison, G. (2007). Interview with Yun Zhang on 3 October at EDAW Brisbane' office.

Keswick, M. (1986). *The Chinese garden.* New York, London, St: Martin's Press; Academy Editions.

RDUDP. (2002). Concept Planning Design of Wuli Lake Area of Wuxi City.

Sun, Z., & Liu, J. et al., Eds. (2007). *Urban Science Research of Wuxi: Li Lake Special* Wuxi, Editorial board of Wuxi urban science research.

Winsor, M. (2002, 20 Mar). DEEP Fusion Project in Wuxi, China: Interdisciplinary collaboration of HK, BRIS, SF offices. Retrieved 3 Oct, 2007, from http://ed.edaw.com/firm/fwCommunications/eNews/eNewsArticle.aspx?aId=24187.

Wuxi Municipal Committee of the Planning and Construction of the Li Lake area, Ed. (2007). *Unveiling the beauty of Li Lake: Wuxi well on its way towards ecological well being* Phoenix Press Group and Phoenix Publishing House.

Wulff, R. (2007). Interview with Yun Zhang on 25 September, at Tract consultants' Melbourne office.

Wulff, R. (2010). Communication through telephone, called to University of Melbourne, on 12 May 2010.

Chapter 9
"Glocalization" through International Practice in Landscape Architecture

Abstract Chapter Nine has three main sections. Section One positions landscape architectural practice in an international context through a discussion of local demand to accommodate ecological recovery and the tourism economy; the role of landscape architects in hybrid international teams; the impact of international landscape architectural practice on the local profession; and ways to bridge local and global transfer. Section Two argues for genuine engagement with sites and diverse local stakeholders through a discussion of Australian landscape architects' approach to gaining knowledge of the Li Lake area, three categories of the identified characteristics, and, the localisation of international design practice. Section Three discusses Australian landscape architects' difficulty in gaining implicit knowledge of place, including ways to convey implicit knowledge to them and their accessibility to the local communities.

Keywords Hybrid teams · Genuine engagement · Implicit knowledge · Local profession · Local community

9.1 Positioning Landscape Architectural Practice in an International Context

This research argues for a more relevant role of landscape architecture in the context of international planning and design practice, whereby landscape architects act as consensus builders. These hypotheses are grounded in theory concerning the urban practice of landscape architecture and the condition of Asia's urbanisation (which has overtaken that of Western countries in terms of scale and speed). On the one hand, urban practice has become hybridised in a way that requires the integration of various creative disciplines. On the other hand, Asia's greater urbanisation has created a market for Western design practice in Asian countries. Theory suggests that the new demand for urban practice can be met by landscape architects' expertise in providing ecological care, and by their multidisciplinary skills in acting as consensus builders. This research identifies four areas in which landscape architecture has developed a key role in the context of the research embodied in this section.

Y. Zhang, *Place Making in International Practice of Landscape Architecture*, Advanced Topics in Science and Technology in China 64,
https://doi.org/10.1007/978-981-16-2442-1_9

9.1.1 Ecological Recovery and Tourism

In the international context, landscape architectural practice is driven by local demand for two types of services. First, as the case study showed, Australian landscape architects were required to provide planning and design solutions to accommodate the recovery of the local ecology of the Li Lake area in China. As reviewed in Chapter Two and Chapter Five, urbanisation underpins the development of local areas and the infusion of diverse international planning and design practices in China. Likewise, the survey showed that Australian landscape architectural services in China are dominated by this kind of urbanisation project. This research further suggests that the demand for the recovery of local ecology and the promotion of the tourism economy have placed landscape architectural practice at the frontier of the international market of planning and design practice.

Landscape architects achieved this prominence through skills in both functional planning (principally for ecological recovery) and aesthetics (principally for tourism economy). First, in terms of ecological recovery, as found in the case study, the two international design teams and Australian landscape architects first became involved *after* scientific teams set the criteria for the ecological recovery of Li Lake. The case study showed that the water treatment of Li Lake was a demonstration project of the 'National 863' environmental recovery project of Lake Tai. The Chinese central government financially supported the water quality improvement of Li Lake. Also, seven Chinese research institutes and universities offered technical support, through which the causes of the pollution were identified, and goals and proposals were set (Nian 2007: 22). The problems identified by these scientific teams included that ninety percent of the water edge of Li Lake was laid with stones and was too steep to allow shallow water for water plants (2007: 22). All of these suggest that clients' expectation of international landscape architectural practice was not aimed directly at the expertise in environmental science. This seems to contrast with the comments made by clients, who said that 'the expertise of foreign landscape architectural firms to build ecological environment can be top-ranked' (EBOCWJ 2002). It could be argued that such an expertise was contributed by the international design teams in three ways, all relating to the work of functional planning. These include first, zoning the active and passive uses of the waterfront land; second, defining diverse treatments of the lake edge; and third, proposing ecological islands and wetlands in the lake.

The above three ways were shown in a number of ways. The RDUDP team proposed passive uses for most of the Li Lake waterfront along the Turtle Head Peninsula (a traditionally popular scenic area), and active uses for the land bordering the proposed new city. Likewise, the EDAW team proposed soft lake edges for passive park use and the newly established water plants; and proposed hard lake edges for active park use and intensive urban use of the proposed new city. Additionally, ecological islands for passive use were proposed in the original fish ponds area. The proposals made by the RDUDP and the EDAW team suggest that specifically, the local demand for landscape architects' environmental expertise was to define a balance between new urban use and the recovery of the native landscape.

It could be argued that in clients' decision to invite foreign landscape architectural firms, such a specific demand had not been explicitly defined. Instead, local government clients regarded environmental expertise of foreign landscape architects as enhancing the natural characteristics of the Li Lake area. This, however, resulted in the three approaches to ecological recovery being little connected to the cultural meanings implied in local activities and ways of deploying native species. Such a disconnection was shown in the mere allocation of activities as active or passive; water edge treatments as soft and hard; and emphasis on local natural characteristics. The reasons for this disconnection are further discussed in Section Two.

As discussed in Chapter Two, the profession of landscape architecture values hidden aspects of place, such as meanings associated with physical objects. In their work for local ecological recovery, landscape architects should investigate sites, engage with the community and conduct research to identify meanings associated with living and dynamic elements, such as the earth, rivers and plants. This is the unique approach of landscape architecture to place and practice. It also means that landscape architectural approaches should creatively express the outcomes of the various scientific recommendations for local ecological recovery. In the context of international landscape architectural practice, the provision of environmental expertise thus also requires the identification of meanings associated with the natural components of place and imposition of these meanings on the scientific recommendations for ecological recovery.

Second, the case study showed that landscape architects were required to beautify and improve the image of the local area to promote the local tourism economy, stemming largely from the local government's awareness of international trends in tourism. This approach was manifest in three ways. First, Wuxi's urban development aimed to brand Lake Tai as the trademark of the local tourism economy (Sun and Shao 2007). Second, the local government expected to use Western concepts and principles of planning and design to improve the image of Wuxi city (Sun et al. 2007). Third, the local governmental developers announced that the successful design of Hong Kong Disney Land was one of the reasons it selected the EDAW team to take part in the case project (Wuxi Government 2004). As Chapter Six reviewed, such local demand was underpinned by the intention to sustain local economic growth in the face of competition with nearby places and to win global recognition (Sun et al. 2007). This aim was also evident in the use of world famous harbour cities, such as Sydney and Geneva, as benchmarks by local governmental clients (2007).

The local goal to promote the tourism economy emphasised building an appropriate image of the Li Lake area. It could be argued that such an emphasis promoted the infusion of global ingredients with preconceived notions of what international design practitioners would bring with them, such as accommodating tourists' activity and establishing landmark architectural forms or landscape statements for the image building of the Li Lake area. The case study showed that the latter two have been materialised in the Western Li Lake waterfront, in the form of the 120 m high lake fountain, and in the architectural forms of the shelters of the boardwalk and amphitheatre. This, however, suggests that tourism often won over ecology and communal values in the process of development.

9.1.2 The Role of Landscape Architects in Hybrid International Teams

Theory (Arvidson and Martin 2007) suggests that landscape architects act as consensus builders among various stakeholders including developers, institutions, local communities and professionals. Such a role of consensus builder is supported by the multidisciplinary skills of landscape architects and suggests the leadership role played by landscape architects. Theory (Wall 1999; Corner 2006) also suggests that landscape architecture has had a more relevant role in urban practice that landscape architects not only contribute their multidisciplinary skills, but also their knowledge of ecology. The case study supported the definition of landscape architecture as becoming a crucial profession in the system of built environmental professions. However, this research highlighted two nuanced aspects of the role of landscape architects as consensus builders.

First, landscape architects take on leadership roles in the complex professional negotiation that delivers urban form. In the international context, such complex professional negotiation is evident in the composition of international design teams with a range of professionals from different disciplines and from geographically distant firms or organisations. It is also evident in the holistic provision of services from planning to landscape design, design development, and delivery and construction supervision. As Chapter Two analysed, a 'more relevant and active role' of landscape architecture in urban practice requires 'the synthesis of landscape, architectural and urbanistic skills into a hybrid form of practice' (Wall 1999: 247). In the context of international design practice, ways of integrating skills and forming hybrid forms of practice were achieved through consortiums of environmental professionals and the leadership role performed by landscape architects. In this study, Australian landscape architects brought together urban planners, urban designers, architects, graphic designers, economic planners, and experts in ecology.

The principal role of landscape architects was also evident in the way that local government clients referred to the hybrid international consultancy teams as landscape architectural firms. Furthermore, the cases showed that among diverse disciplinary experts, landscape architects directly engaged the top local stakeholders—the local government developers. It can be argued that the presence of landscape architects in the hybrid international design teams revealed the power of decision-making in providing a suite of services. However, these findings contrast with the fact that the business names of these Australian landscape architects were concealed under another business name, and even another nationality.

Second, the role of landscape architects as consensus builders in an international context differs from that in the local or national context of design practice. The cross-cultural conditions of international design practice, however, reduce the power of consensus-building. As reviewed in Chapter Two, in the controversy of changing places, landscape architects can build a common language among diverse stakeholder groups with their multidisciplinary skills (Arvidson and Martin 2007). Chapter Two also revealed that this role for landscape architects is more relevant in urban practice because of the new demand for integrated skills from diverse professions including urban planning, urban design, architecture, art and engineering. Due to the often organic or unrecordable aspects of landscape design, such as the positioning of a rock according to its shape, landscape architects often interact closely with the construction team and even modify a design on-site. Therefore, landscape architects' engagement with a range of stakeholder groups is consistent with that of a consensus builder.

In the international context of planning and design practice, it is possible to identify groups of stakeholders. In the case study, two cultural groups emerged: stakeholders from the practitioners' culture and those from the host culture where the services are provided. The contributions of different professionals from Brisbane, Australia, to the case project demonstrated the presence of the cultural group of practitioners' own. Figure 9.1 shows how a boundary became evident between the two cultural groups. This boundary reduced international practitioners' access to a range of local stakeholder groups and reduced the extent of consensus. Theory (Hough 1990; Armstrong 1994; Arvidson and Martin 2007) suggests that in the case of local design practice, landscape architects have access to diverse stakeholders (Fig. 9.1: top). By contrast,

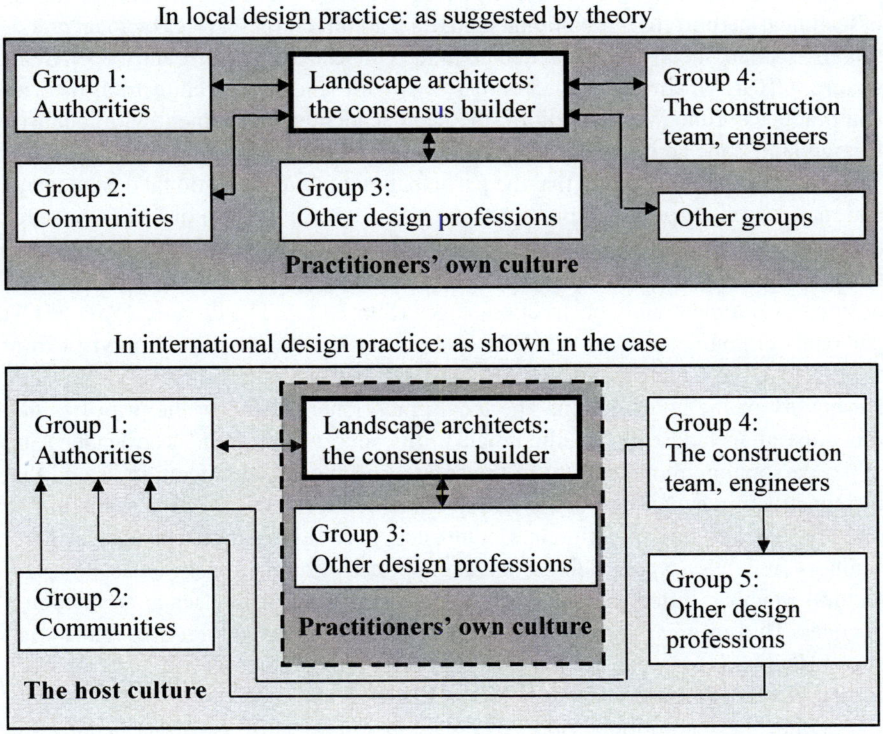

Fig. 9.1 Landscape architects' role as consensus builders. *Source* Top–adapted from theory (Arvidson and Martin 2007), bottom–the case study

in the context of international planning and design practice, as this research showed (Fig. 9.1: bottom), communication with other local stakeholders was blocked by this cultural boundary (except with government clients). A non-cross-cultural process emerged in the cross-cultural design practice.

The case study also identified that diverse professionals from the practitioners' culture were mainly involved in the planning phase. By contrast, local design professionals were mainly involved in the design development phase, when construction and changes to the designs began. At that time, local design professionals were in the position to provide technical support for landscape construction. The later involvement of local design professionals further reduced the power of consensus-building in international design practice.

9.1.3 The Impact on the Local Profession

International practice in landscape architecture has impact on the local profession in China. This was shown in the development of landscape architecture as a modern profession there, which was influenced by the West in three ways: first, the provision of technical support for international landscape architectural services by local practitioners; second, the professional debate in the Chinese Society of Landscape Architecture (CHSLA), stirred by the comparisons with landscape architectural theories and practices of the West; and third, the local academic engagement with localised foreign landscape architectural firms.

First, the case study found that the participation of two international design teams and major services provided promoted the reflection and growth of local landscape architects. The case study showed that the local practitioners witnessed the advanced foreign landscape architectural services in their support for the EDAW team in the design development and construction stages (Sun 2007). The director (Yang 2007: personal communication) of WMCPC also praised the construction supervisor from the EDAW team, saying that this Australian landscape architect effectively controlled the quality of the construction. The local practitioners also saw the priorities that this Australian landscape architect had in his supervision work. The leader (Sun 2007: personal communication) of the supporting local design team indicated that the construction supervisor of the EDAW team could report problems to the mayor directly, but the local practitioners could not. It could be argued that local practitioners' reflection upon their technical support for international landscape architectural practices lifted the standard of local provision of landscape architectural services. This progress is evident in the increase of local design practices in the other sites of the Li Lake waterfront, which was after the completion of the major work for the Western Li Lake waterfront by the EDAW team.

Second, the institutionalisation of landscape architecture in China occurred in parallel with the flourishing of international landscape architectural practice in China. As Chapter Five discussed, there was a nine-year delay (1998–2006) in the establishment of a registration system of landscape architects in China. During that period, in addition to the infusion of foreign landscape architectural practice in China, Western theories and projects in landscape architecture were increasingly introduced into China. Chapter Five also found that the debate in CHSLA in this period was driven by comparisons with the West. This debate peaked in the early 2000s, when the role of planning in landscape architecture and ways to respond to local senses of place were two major topics.

As the case study shows, Chinese clients' recognition of the hybrid international design teams as landscape architectural firms suggests that, at the beginning of the twenty-first century, planning for vast land change was recognised and conceptualised as landscape architecture. As Chapter Five showed, China's demand for international planning and design services covered various professions, such as architectural design, urban design, city planning, and landscape design. International landscape architectural practice was always a part of these professions, concealed in firms that were named after world-famous architects and recognised mainly in residential landscape design until the late 1990s. The demand for international landscape architectural practice in China differs from the demand for international architectural and planning practices, which initially focused on fashionable Western architectural forms or economically-focused urban morphologies.

Chapter Five showed that around the year 2000, planning practices in China tended towards being environmentally and culturally sensitive, although the latter has not yet been a success. The case study revealed the relevance of international landscape architectural practice to the four new planning practices in China: conceptual planning, design-based planning, environmentally sensitive planning, and culturally sensitive planning. This was first evident in a range of titles used for the principal planning documents of the case project. These titles included Concept Planning Design, Concept Master Plan, and Conceptual Plan, which were used by both the clients and international design teams. The planning also included the design of bridges and precincts, further suggesting it was design-based planning. While international planning practice was often in the form of conceptual and design based planning, international landscape architectural practice, by contrast, also incorporates environmentally sensitive planning. In relation to the major debate in CHSLA on the role of planning in landscape architecture in China, the form of environmentally sensitive planning, as shown in the case, offered practical examples to the local profession through land use zoning, lake edge treatment, and enhancing natural characteristics.

In terms of culturally sensitive practice, the boom of international landscape architectural practice in residential sectors in China, as identified in Chapter Five, shows the demand for new forms of outdoor spaces to accommodate socio-cultural change in China. This finding suggests the relevance of international landscape architectural practice to culturally sensitive design practice. However, in terms of planning, there is a gap between the local demand for the cultural construction of place and the approach of international design teams in the planning phase. This gap consequently influenced the subsequent landscape design. In relation to another major debate in CHSLA: ways to sustain the characteristics of place, this gap suggests that the practical Western approaches have yet to be introduced in China. This gap is further discussed in Section Three.

Finally, the increasing interaction between foreign landscape architectural firms and Chinese professional journals and academia, all demonstrate the impact on local professionalisation. Such interactions further suggest the focus in CHLSA on environmental expertise. Chapter Five found that in 2004, the international firm EDAW became the first foreign firm to introduce its projects in the main professional journal for landscape architecture in China, where CHSLA's selection of their projects were all about waterfronts that had been affected by land development. Furthermore, there have been symposia on the topic of ecology that were jointly held by the EDAW firm and universities in China, such as Beijing University and Tongji University in 2005, and Qinghua University in 2006 (Fu 2007). These observations suggest that landscape architects who work internationally have been engaged in disseminating up-to-date and practical experience through their interaction with the educational system for the local professions. The concerns with waterfront areas and ecology by CHSLA, again suggests the increasing emphasis on environmental expertise for land development in China, and the consequent request of such expertise from overseas and from Australian landscape architects.

It could be argued that the exchange of knowledge through the abovementioned approaches centred around knowledge of ecology. It could even be argued that in the relatively new market condition of providing planning and design services in China, landscape architects should play a more active role through their work internationally, such as improving clients' understanding of intellectual property in planning and design and enabling local practitioners to gain confidence in the new market conditions. Such a role is articulated in AILA's code of professional conduct (AILA 2005). Specific to professional colleagues, AILA's code of ethics requires its members to 'notify' their professional colleagues before accepting the commission from a client, when their colleagues are known to have undertaken the work. The code also shows explicitly the need to be 'in conjunction with a local colleague when undertaking work in a foreign country' to 'ensure local culture and place are recognized'. These requirements suggest that to play such a role, landscape architects working internationally should enhance the collaboration amongst themselves and engage more closely with local practitioners and pursue joint practice.

9.1.4 As a Means to Bridge Local and Global Transfer

International landscape architectural practice contributes to global connections. As reviewed in Chapter Two, contemporary places are embedded in the constant interconnection between economics, culture and ecology. This is evident in the local demand for enhancing the tourism economy and the consequent use of would famous harbour cities as benchmarks, as analysed in the case study. Four approaches that contribute to this global connection are identified in the case. These approaches also suggest that international landscape architectural practice could better act as means to bridge local and global transfer.

One approach is the use of popular western landscape forms to accommodate lasting local activities. The case study identified the use of place images from around the world by international design teams. The approach of the RDUDP team suggests a focus on the functions and aesthetic meanings associated with water features and the consequent landscape forms. Physical forms presented by these images were not further applied in their planning and design. The approach of the EDAW team suggests a focus on the histories of waterfront renovation and relevant strategies, and the everyday commercial experience in Western situations. This suggests the major purpose of seeking applicable similarities between the Li Lake area and the developed West in using these images, rather than directly copying their physical forms alone in applications on site.

The different methods for using images employed by the two teams suggest varying approaches to bridge local and global transfer. One is the use of foreign forms for local activities, such as commerce and recreation. The other is the use of local forms to accommodate activities of new comers from worldwide. The former is evident in the introduction of the artificial sandy beach. The sandy beach was not a natural character of the local area, yet it is quite usual in Australia. The proposal of the sandy beach was first shown in the structure plans provided by the RDUDP team. Following the plan and landscape design of the EDAW team, the sandy beach was constructed, using sand transported from other areas (Brookbank 2007: personal communication). The case study found that the beach has been a popular place for local residents of all ages, especially in hot seasons. It promotes the activity of engaging with water, although swimming was prohibited during that time, as the water condition remained under recovery.

The example of the sandy beach suggests that key to the successful use of popular Western landscape forms is the identification of popular local activities. It allows more space for the localisation of Western forms, as local life is more likely to unfold there, and thus a sense of place can be cultivated. In contrast, the use of local forms to accommodate international commercial and recreational activities was not evident in the case study. One reason is that the new city, which is aimed at attracting investment and business from outside Wuxi and China, remains under design and construction. This suggests the extensive time needed for enabling a local place to accommodate activities from the rest of the world, such as multinational business including tourism. The popularity of the imported form of the sandy beach also contrasts with the empty paved plazas, which are often popular in the West, where gathering, exhibition, sun bathing and performance are vividly demonstrated. The leader (Sun 2007: personal communication) of the local design team indicated that a coffee house designed by the EDAW team, was demolished soon after it was built. This is an unsuccessful example of imported functions and forms. On the one hand, it did not service established local activities. On the other, the new urban life and international commercial activities it proposed were yet to be developed. This example, in turn, suggests that using Western forms to accommodate local activities that will remain popular in the long-term is a more effective way to bridge local and global transfer.

The second approach was the clients' tour of famous waterfront cities in Australia, such as the one in Brisbane led by an Australian landscape architect from EDAW. This is also an effective way of introducing applicable Western experience. Compared with sightseeing, such an approach allows non-Western clients to develop a deeper understanding of places, which were used by them as a benchmark for local development. The explanation of those places to clients was offered by those who understood or participated in their evolution, it thus reduces the risk of merely using the physical form as a benchmark for development.

The third approach was international practitioners using new ways to express locality visually. Tuan (1977: 159) suggested that familiarity and a sense of attachment is grounded through an 'undramatic tie to locality' and 'lack of sharpness'. Tuan's words explain the fact that the experience of place may be difficult to articulate by those who are too familiar with it. This in turn suggests that local clients expected new ideas from international designers, whose creativity can be inspired by their unfamiliarity with the local place. As identified in the case study, new ways of expressing the local familiar were evident in architectural forms suggesting sails and lake birds. Using Tuan's (1977) words, this is the introduction of a dramatic tie to locality. However, the case study showed that this approach suggests an emphasis on representing visual local elements alone as opposed to an array of sensorial experiences sourced form the locality. This is further discussed in Section Two.

The fourth approach is the use of local materials and construction skills to localise forms and technologies that are conventional in the West. The case study revealed that this approach was adopted passively, mainly under the limitation of the availability of plant species, hard landscape materials, and local manual operation skills. For example, Corten steels were previously proposed for the edge material of an undulating lawn area. Such materials are also popularly used in the urban landscapes in Australia. These materials were replaced by stainless steel in construction (Fig. 9.2), simply because they were not locally available and therefore too expen-

Fig. 9.2 Materials used in the case site. *Source* EDAW's 2005 waterfront awards application package, courtesy of Brookbank

sive. The local design team leader (Sun 2007: personal communication) indicated that the team understood that the corten material would be visually preferable, yet, it was not allowed by the project budget.

The case study also identified that the use of manual rather than machine operation were seen by international practitioners as affecting quality outcomes. Figure 9.3 shows local artisanship in the traditional Li Garden, which is nearby the site of the case project and the village of Double Bay, which has been conceptually planned by the RDUDP team. The figure shows that the different colours of cobbles compose the physical form of a fish in a path of the Li Garden (Fig. 9.3), which contrasts with the precast paving of a fish in the case site (Fig. 9.2). It could be argued the integration of local materials and artisanship into popular Western landscape forms, rather than using them as an expedient measure, is an approach to bridge local and global transfer.

Frampton's (1983) critical regionalism suggests a dialectic approach that on the one hand uses modern technology and on the other hand commits to local specificities. Such an approach is manifest in the water-board shelter, in which new technologies and modern materials are applied and the architectural form suggests the traditional form of local sails. This example shows the approach of indirect borrowing in architectural terms, more powerful than the direct use of the physical forms of fish, birds and historical personages, which can be seen as a form of tokenism. It could be argued that the sandy beach shows a further indirect borrowing, as this approach related to local behavioural character rather than physical form alone. This concern is most relevant to Schulz's (1980) argument for genius loci and a phenomenological approach to architecture. It could even be argued that a phenomenological approach to landscape architecture is promising as a bridge for local and global transfer in the international context, where landscape architects are at the frontier of the conflicts between contemporary universal civilisation and local specificities.

Fig. 9.3 Local artisanship shown in other sites. *Source* Author, photographed in November 2007 in areas near the case site

9.2 Levels of Genuine Engagement with Sites and Stakeholders

This research argues for international practitioners' genuine engagement with sites and diverse local stakeholders to gain explicit and implicit knowledge of place. The survey and case study revealed three major approaches to gaining such knowledge in the context of international planning and design practice. These include the acquirement of initial understanding through learning from collective knowledge; practitioners' visits and study of the physical sites and nearby areas; and their formal and informal engagement with local stakeholders. The research suggests that ways to engage with sites and local stakeholders influenced Australian landscape architects' gaining of implicit knowledge of the Li Lake area. Such influence is reflected in the three categorise of local characteristics that the two international design teams used as basis for their planning and design. The research also suggests that along with the localisation of the practice, international practitioners' understanding of local characteristics extends from the heroic part of ancient local history to a more subtle local fishery culture.

9.2.1 Approaches to Gaining Local Knowledge

In the context of international design practice, the first approach to gaining local knowledge is through learning from collective knowledge. Collective knowledge means knowledge established from previous international design practice and knowledge that can be publicly accessed (e.g., through the Internet and publication). This approach allows international practitioners to understand, first, how to operate a practice in a different socio-cultural context and, second, the general information of sites or places that have been proposed for change.

As Chapter Two showed, strategies for non-local designer's landscape investigation and design include four steps: landing, grounding, finding and founding (Girot 1999). According to Girot (1999), the first step, landing, is a personal experience of moving from the unknown to the known. It involves reactions to the difference between the preconceived and reality and consists of feeling and knowing difference. Thus, landing, the primary stage of design practice is influential in forming initial design responses. Learning from collective knowledge in international design practice, however, contrasts with this individual experience of landing.

The case study demonstrated the collectiveness of landing in a number of ways. The RDUDP and EDAW teams and their key members already had knowledge of work in China before they became involved in the case project. As discussed in Chapter Six, international design practice in China was initially specific to ethnic areas and Hong Kong was a major stepping stone for Western design firms to have offices and work in mainland China (e.g., the Hong Kong practice of Tract Consultants in the 1970s). The case study also showed that planners and designers from

Tract Consultants played a key role in the RDUDP team by providing planning services. In addition, the leading firm of the RDUDP team, Ron Dennis, had a previous business relationship with Tract Consultants in Australia and had provided a range of consulting services, from farming to planning, in China (Wulff 2007: personal communication).

A further precedent for work in China came from EDAW Asia in Hong Kong. The firm was established in 1997 when the international firm EDAW, which is based in the US, merged with the British firm Earth Asia, originally in Hong Kong. As the case study showed, the EDAW Brisbane office received a brief through the Hong Kong office and provided the technical response, and the Hong Kong office gave the business response (Harvison 2010: personal communication). In addition, early in 1998, EDAW had just completed its services for the development of Jing-ji Lake in Suzhou, China, in which the principal of the EDAW Brisbane office helped to establish the EDAW Suzhou office (Harvison 2007: personal communication). Suzhou is about fifty kilometres from the Li Lake area. These findings indicate that the EDAW team had an established knowledge of the region prior to providing services for the case project.

Furthermore, using established knowledge for landing seems to be essential for international design practice. Chapter Five found that first-hand and long-term knowledge of the Chinese market is critical to Australia's business in China. The APSFADTRC and Forshaw (1996) encouraged service providers to capitalise on the expertise and experience of staff returning to Australia after postings in China. Their suggestion indicates the essential need to transform individual experience into collective knowledge for subsequent Australian practice in China.

The case study showed that, compared to the EDAW team, the RDUDP team's key planners appeared to have less direct access to first-hand knowledge and experience because they were contracted under another firm, Ron Dennis. By contrast, although the EDAW Hong Kong and Brisbane offices provided business and professional responses, respectively, EDAW established a platform under its wider umbrella. Thus, its team had more direct access to first-hand knowledge and experience. This access was also evidenced by the e-news issued by EDAW's San Francisco and Hong Kong offices, which reported on the case project's progress when the Brisbane office was conducting the major work of planning and design. The Australian landscape architect (Brookbank 2007: personal communication), who acted as the construction supervisor of the EDAW team, also indicated that before he joined the team and started to work for the Shanghai office, he studied relevant documents in Australia and obtained relevant knowledge of the Li Lake area from his colleagues both in Australia and China.

The collective approach of landing was also demonstrated by the fact that both the RDUDP and EDAW teams were staffed by professionals from offices in geographically different areas or from different firms. Thus, the teams were composed based on collaboration and knowledge sharing. The survey also found that existing professional links are one of the three major sources for Australian landscape architectural businesses to obtain work in China, which again suggests a collective approach for landing. This collective landing in international design practice seems increasingly widespread in ongoing international projects and in the increased efficiency that computer skills and rapid transportation make possible. In addition, technology (e.g., digital land information systems and satellite imagery) provide professional staff with extensive data for the sites where services are to be provided before they conduct on-site investigations.

The collective approach to landing suggests that technical skills are crucial in the initial stage of international design practice. This is apparent in using information technology to communicate between geographically distanced professional groups, retrieving publicly available information from the internet, and referencing knowledge and experience accumulated through previous practices. These technical skills are undoubtedly necessary, especially for international projects that are often large in scale, complex in operation, and situated in socio-cultural contexts that differ from the practitioners' own. However, dependence on technical skills and collective approaches in international design practice inevitably reduces the designers' individual experience of a place. Although landing may consist of both collective and individual aspects in international design practice, the collective approach appears to be the more pervasive of the two. It could be argued that a central task of the collective approach to landing is to support a more genuine engagement in the subsequent stage: in Girot's words, grounding. Figure 9.4 shows how the collective approach to landing should provide contextual information before an individual approach to landing and provide strategies for a more genuine engagement in grounding.

The second approach to gaining knowledge of the Li Lake area is international practitioners' visits and study of the site and nearby areas. As the case study showed, the Australian landscape architects, as key players of the international design teams, undertook the second approach through intensive site investigation and periodic visits. Obviously, on-site investigation is fundamental to landscape architectural practice, whether local or international. However, in the context of international design practice, the process of investigating a site seems more intensive because, in the early stage of practice, international practitioners often have a limited time to

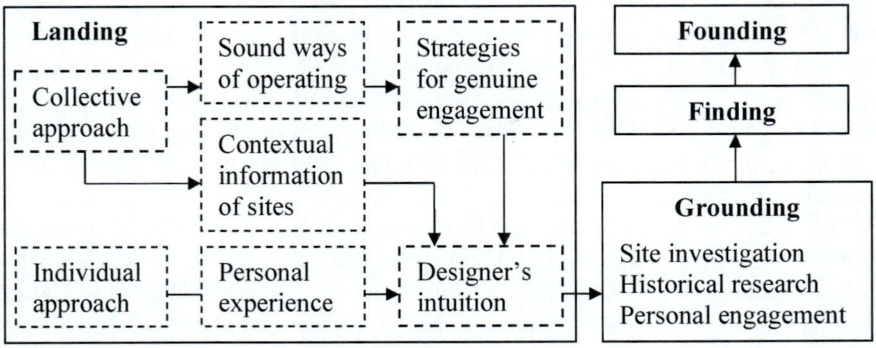

Fig. 9.4 Exploring place quality in international design practice. *Source* Adapted from the case study and the theory of Girot (1999) and Hough (1990)

generate concepts and are less likely to work locally on a long-term basis. Moreover, localised international practice usually occurs after the planning and design schemes have won the subsequent contract. This process was evident in the transferral of the principal work from the EDAW Brisbane office to the Shanghai office after EDAW's planning scheme was selected and the landscape design of the northern shore of Li Lake was developed. The survey also found that repeated visits were used more in pre-design phase than in the schematic design phase. Therefore, in the early stage of international design practice, the process of landing to gain individual experience and the process of grounding for genuine engagement are relatively short. Furthermore, the switching of key international practitioners at different stages in the project further fragmented the already narrow range of individual experience that is crucial for outside designers to form design intuitions.

The case study also identified two methods for international practitioners' intensive site investigation. One was site visits that were supported by WPB; the other was practitioners' self-investigation. The director of master planning (Wulff 2007, 2010: personal communication) of the RDUDP team indicated that a key planner on the team carried out detailed visits to the local area and identified valuable aspects of local places that were not included in the background information that the clients provided. The case study also revealed that both international consultancy teams conducted studies to identify valuable views and visual corridors to connect the existing historic landscape to the proposed new city. These findings suggest the importance of self-guided investigation, which can uncover perspectives beyond the clients' focus. However, they also show the emphasis on the visible aspects of place, rather than on both the visible and hidden, as Girot (1999) suggests. It could be argued that insufficient knowledge of hidden aspects of the Li Lake area was a result of the short-term locally based work in planning phase.

Intensive site investigation and repeated visits are a genuine engagement with the physical site. Although this approach can identify the visually prominent local elements and gain explicit knowledge of the Li Lake area, it does not provide sufficient implicit local knowledge. As Girot (1999: 62) showed, grounding is the discovery of a site's visible and invisible significance; therefore, grounding is not achieved only through repeated visits and studies, but 'has to do with orientation and rootedness'. As Relph (1976) argued, the identification of orientation and rootedness of place is the third level of experiencing the 'insideness' of place (Relph 1976: 142), which is the process of a responsible outside designer identifying a place's community-based values. As Hough suggested, such values are revealed through planners and designers' interaction with local communities These insights suggest the importance of hidden aspects of place, which should be identified in the early stages of planning and design practice, and the need for three types of work, including first, on-site investigation; second, historical research; and third, personal engagement. The latter is also a genuine engagement with place and manifest in the case study.

The case study found that planners and designers' personal engagement was characterised by direct engagement with the top local stakeholder—the government developer, but a less direct engagement with the other three major stakeholder groups (Fig. 9.5). The other three groups included local experts, local communities, and local

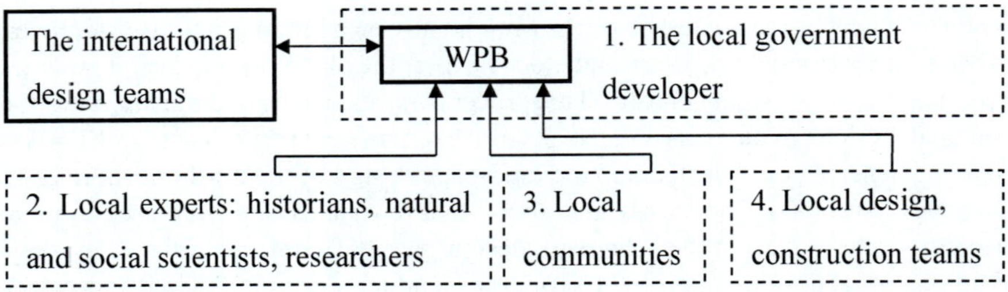

Fig. 9.5 Engagement with the four major local stakeholder groups. *Source* The case study

design and construction teams. The direct engagement with the top local stakeholders was explained by the role of Wuxi Planning Bureau (WPB), a key representative of the government, in a number of ways. WPB not only delivered official documents and messages from the government to the international consultancy teams, but also organised both formal (organised meetings) and informal (banquet) engagements. WPB acted as a communicator between the international design teams and the other local stakeholders. Through WPB, the international design teams communicated with local experts and representatives from other government departments. The director (Wulff 2010: personal communication) of master planning of the RDUDP team explained that during the primary stage of the planning phase, the minister of WPB came every day to take queries and suggestions from the team and brought answers and comments from local experts back to them. Further evidence of the closeness of the relationship with the government lies in the fact that the website of EDAW lists WPB as one of their many local partnerships in China.

The teams' direct engagement with WPB operated in stark contrast to its less direct engagement with the other three local stakeholder groups. Moreover, the sequence of the team's engagement with the four groups of local stakeholders differed noticeably. The case study showed three major phases of the international design practice: planning, landscape design and landscape design development. The teams' direct engagement with WPB and less direct engagement with local experts continued throughout the three phases. However, the team's engagement with local communities followed the planning phase. Furthermore, the team's engagement with the local design and construction teams started in the landscape design development phase, thus there was a distinct discontinuity. The partnership that the team built with the local government developer through WPB was also a genuine engagement with local stakeholders. However, it could be argued that international practitioners' indirect and late engagement with other local stakeholders reduced the likelihood of their acquirement of sufficient knowledge, especially implicit knowledge, to establish an understanding of what constituted a sense of place. These issues are discussed in the following sections.

9.2.2 Categories of the Identified Local Characteristics

This research argues that the identified local characteristics, as the case study showed, could be classified in three categories: the visual vernacular, the heroic past, and lasting local activities. These three categories reflected the existence of genuine engagement in the context of international design practice.

The visual vernacular refers to the visually prominent local elements. One area of the visual vernacular, as was interpreted in the planning reports of the international design teams, is the traditional architectural forms that characterise Jiang-nan. As Chapter Seven discussed, these include traditional garden elements, buildings, bridges and water villages. As the case study showed, these traditional architectural forms were major historical references in the planning report that the RDUDP team produced. Likewise, these forms were cited in the 2005 waterfront awards application package by the EDAW team. Figure 9.6 shows the traditional architectural forms that have been constructed on the present Li Lake waterfront. The EDAW team described these forms as referring to a 'traditional Wuxi style bridge' and 'traditional Chinese shade walk' (EDAW 2005).

The other area of the visual vernacular, as the case study showed, was the unique natural environment of the Li Lake area, including the vast scale of Lake Tai and its sense of wildness. Hough (1990) argued that the bottom line to retaining the identity of a regional landscape is to maintain the area's basic topographic and biological character. The case study showed that such a bottom line was considered in the planning and design work of the international design teams, through integrating the ecological recovery of Li Lake that was already underway. The RDUDP team's proposal of active and passive land use and environmental island, and the EDAW teams' strategies for soft lake edge, wetland boardwalk were all relevant to the enhancement of the natural characteristics of the area.

In the constructed waterfront, a range of areas are planted with native water plants and with boardwalks passing through them. In these particular places, the immense body of water in the far distance further reminds the sense of the original natural

Fig. 9.6 The visual vernacular in traditional architectural forms. *Source* EDAW's 2005 waterfront awards application package, courtesy of Brookbank

Fig. 9.7 The visual vernacular in local wilderness. *Source* Left—*Unveiling the Beauty of Li Lake: Wuxi Well on its Way towards Ecological Well being* (WMCPC 2007: 30), right—author, photographed in November 2007

areas of the Li Lake waterfront that are relatively wild. Figure 9.7 shows such a sense of place in the ecological islands that have been established in the originally polluted fish pond area. Chapter Six discussed how this sense of naturalness distinguishes the Li Lake area from nearby cities such as Suzhou, which is characterised by a more delicate landscape. This aspect of the visual vernacular at the area is valuable in promoting the value of nature conservation in the on-going urbanisation and management of the waterfront.

The case study showed that the identified visual vernacular also included the physical forms of fish, waterfowls, fishing boats and other fishing instruments that suggest the local fishing industry. The case study found that these forms have been represented in paving, sculpture, outdoor facilities and the architectural forms of shelters. The latter have been highlighted as the most successful interpretation of the local characteristics of place (Fig. 9.8). The recognition of the architectural forms of the shelters is evident in the frequent references to them, including in the EDAW team’s international awards application packages, on the websites of the international organisations that issue awards, and in the publications of local government

Fig. 9.8 The visual vernacular in lake sails and birds. *Source* EDAW’s 2005 waterfront awards application package, courtesy of Brookbank

developers. In the EDAW team's 2005 waterfront awards application package, the architectural forms of these shelters are described as 'drawing influence from local Taihu lake fishing boats' and were 'inspired by the wings of water birds as they take off' (EDAW 2005).

The above three areas of the visual vernacular—the traditional architectural forms, the unique natural surroundings, and entities signifying local fishing life, are pervasive in the Li Lake area, Wuxi and the Jiang-nan region. Therefore, these visible elements belong to the explicit knowledge of the characteristics of the area. International practitioners can gain such knowledge through the collective approach and intensive site investigations. In the use of collective knowledge, the visual vernacular can be identified from the often readily available written materials, such as background information of the site that is provided by local government clients, documents accumulated through international practitioners' previous practices in similar regions, or other media such as professional publications and even tourist brochures. Such written forms of information are easier to transmit to international practitioners.

The identification of the visual vernacular via the above approaches is useful in landing—the primary stage of international design practice. However, at this level, the practitioners have had no physical interaction with the site or with local stakeholders. As discussed previously, in the use of collective knowledge (e.g., collecting information from databases and using computer systems and publicly available information), technical skills are dominant. It could be argued that the identification of traditional architectural forms, as shown in the case study, was through such an approach, using technical skills, because these forms have been well documented and easily available in professional sources.

It also could be argued that the identification of the relative wild sense of the original waterfront and the ordinary entities showing local fishery life, rely more on practitioners' genuine engagement with the site. Such engagement occurred not only in international practitioners' intensive investigation of the site in the early planning phase, but also in the later phase when the practice was being localised. The director of WMCPC (Yang 2007: personal communication) described how the landscape architect from the EDAW Brisbane office took a few days and finally got the idea for the architectural form of the shelter of an amphitheatre from the flying birds of Lake Tai. The director indicated that the landscape architect sketched the design in a restaurant in Wuxi. This suggests a more locally-based design process, which occurred in the landscape design development phase.

The second category of the identified local characteristics concerns the heroic past in the local history of Wuxi. As the case study found, the characteristics of the Li Lake area that the RDUDP team identified in the planning phase included the ancient Grand Canal and the ancient urban forms that have endured in Wuxi City for centuries. They also included local stories and past events, such as the legend of the ancient local heroes Fan Li and Zhang Bo. Similarly, the landscape design scheme that the EDAW team provided identified local agricultural, fishing and mercantile

Fig. 9.9 The heroic local past in art works. *Source* Top—author, photographed in November 2007 (marble display with multiple interpretative boards telling the legend of Zhang Bo who turned into a dragon to control flood). Bottom—*Unveiling the Beauty of Li Lake: Wuxi Well on its Way towards Ecological Well Being* (WMCPC 2007: 50) (statues and interpretative boards telling local history)

cultures, and the legend of the above-mentioned ancient heroes. Subsequently, these aspects of the heroic local past came to be represented in a range of art works, such as illustrative boards telling local stories and statues of historical figures (Fig. 9.9), as well as in the names given to the waterfront areas.

It could be argued that a heroic local past is also explicit knowledge of the sense of place that resulted from a closer relationship with the local government developer and a less direct engagement with the other local stakeholders. As reviewed in Chapter Three, explicit knowledge can be obtained through formal exchange: the reading and analysis of relevant statistics, diagrams, and texts recording past events and local stories. In the case study, the clients provided relevant information about the Li Lake area, much as the RDUDP team would have in their practices in Australia (Wulff 2007: personal communication). Aspects of the site such as ancient heroes, recent nobles and outstanding events could hardly have been discovered through intensive investigation of the site by the international practitioners. Such information had already been constructed by local authorities and was readily available, or it existed in the form of collective knowledge that can be publicly available. The emphasis on the outstanding components of the local history suggests that the international practitioners were receptors of the narratives provided by local governmental authorities, rather than actively engaged in collecting and formulating the narratives by themselves. In other words, the local governmental authorities identified the heroic local past and transmitted this message to the international practitioners.

Such an approach to learning about the local history influenced the team's interpretation of sense of place. It reduced the use of local history and tradition to milestones of a heroic past, which resulted in the absence of references to the area's more subtle and vivid collective memory. Young (2006) argued that culture is embedded in geographies, societies and histories but that its voice in planning is weak. The case study showed that the voice of culture was not weak in the planning of the Li Lake area, but focused on the outstanding part. Therefore, the problem was about how history could be used subtly in the practical work of planning and design. While this problem may also be applicable to local practitioners working on local sites of their own countries, a relative familiarity with the site could encourage responsive local practitioners to find subtle solutions. In international planning and design practice, despite the good intentions of planners and designers, this problem is amplified because of the reduced accessibility to implicit knowledge covering a specific place. This is further discussed in Section Three.

The third category of the identified local characteristics was in the behavioural. The case study found that the international design teams identified activities such as practicing Tai Chi, drinking tea, swimming and fishing as references to local specificities. In this category, some activities characterise the nation of Chinese people, such as practicing Tai Chi and drinking tea. Others characterise the people of the Jiang-nan region and Wuxi, who are used to interacting often with water. The response to the category of lasting local activities is evident in the man-made sandy beach, which has been locally popular. It also has been explained by the constructed fishing shelters, which are presently located at the originally operational fish farm (Fig. 9.10, left).

It could be argued that knowledge of local activities also has explicit and implicit aspects. This research suggests that the explicit aspect includes types of activities that characterise the nation and local people and the implicit includes the cultural

Fig. 9.10 Places for lasting local activities—paved and wild fishing areas. *Source* Left—EDAW's 2005 waterfront awards application package, courtesy of Brookbank, (a fishing area at the case site); right—*Inhabitant, picturesque scene and artistic conception: landscape design of northern shore of Jincheng Bay* (Guo and Li 2007: 86) (a locally designed fishing area nearby the case site)

preferences that determine where activities occur. For example, the local designers of a fishing area in 2007 for a site on the Eastern Li Lake waterfront identified that an un-developed waterfront area, which is natural in appearance, had been a popular area for local fishing lovers (Guo and Li 2007). Thus, the designers proposed fishing areas surrounded by wild water plants (Fig. 9.10, right). This idea contrasts with the fishing shelters built in the paved area (Fig. 9.10, left). This preference is also evident in the contrast between the empty paved plazas decorated with planting boards and the areas with a natural appearance, which are used more often, in the case site (Fig. 9.11).

Lin and Wang (2005) argued that, traditionally, Chinese people attempt to imitate wild nature in landscape gardening, rather than the cultivated nature such as agricultural fields. Therefore, flower boards and horticultural values were less developed as a part of landscape gardening in Chinese garden history. Furthermore, Chinese people's recreational use of urban space tends to create street conveniences for everyday use, such as eating, drinking tea, selling and buying. In his argument for the essence of tradition, Miao (1989, 1991) pointed out that in the Chinese tradition, traffic spaces serve as public spaces. Therefore, public spaces in China were often in a linear form. Miao (1989, 1991) also indicated that the significant central area is often an empty space that cannot be reached, such as an island in a lake or a mountain peak.

It could be argued that this spatial organisation of public domain in China differs from that in Western societies, where a church is often located in the town centre

Fig. 9.11 The local behavioural in naturalistic and geometric areas. *Source* EDAW's 2005 waterfront awards application package, courtesy of Brookbank

and surrounded by plazas where people come together. Miao's argument suggests that Chinese public life is essentially focused on street activities and a fluid type of behaviour. It also suggests that active public space in Chinese cities may not take the form of plazas, but linear spaces facilitated with street conveniences.

The under-used plazas and more popular natural or naturalistic areas suggest that appropriate urbanised waterfront forms for local activities are yet to be identified. According to Lin and Wang (2005) and Miao (1989, 1991), this knowledge cannot be formulated from the heroic local past or constructed simply from cues taken from the visual vernacular that is gleaned from site investigations. Chapter Six analysed a unique aspect of China: tradition is spiritually active, but physically invisible. It could be argued that such knowledge lies imperceptibly in the history and tradition of the nation and the local people and that it remains not well articulated in the realm of this case study.

9.2.3 The Localisation of International Design Practice

The case study identified an international-oriented conceptual delivery and a local-oriented material delivery of place of international design practice. The former prioritises popular landscape and architectural forms in the West and in Australia, and the latter creates hybridisation. The transformation from the international conceptual delivery to local material delivery was a process of localising international design practice. This transformation was also evident in the survey findings, which revealed that the trajectory of localising Australian landscape architecture practice in China took place in three steps. The first step was the less locally based pre-design services. In the second step, schematic design services initiated the process of localisation by establishing branch offices. The third step took the forms of construction coordination and landscape maintenance as localised practice.

The introducing of popular Western forms through international conceptual delivery suggests the homogenising effect of international design practice. Daniels (2005: 265–282) argued that the process of localisation in people's everyday lives and experiences is a process of cultural hybridisation; the movement of people, objects and images around the world is a process in which cultural homogenisation is often observed. As Lorch and Cole (2003) and Bull et al. (2007) argued, international design practice may act as an agent to promote the flows of Western styles and global homogenising forces. This case study observed the movement of the key personnel from Australian landscape architectural firms and the delivery of images of waterfront places from around the world, including tours for local governmental clients in Australian waterfront cities. These factors suggest the association of international practice with the process of cultural homogenisation. It could be argued that, through such temporary movement, international styles of landscape and architectural forms, and materials and construction methods were infused into the designed and constructed waterfront area.

The phases of landscape design development and the parallel construction supervision were, by contrast, locally-oriented processes. As discussed previously, the availability of materials was restricted by local conditions, and local construction methods influenced the materialisation of the waterfront landscape. Also, the EDAW Shanghai office gradually took over the principal work from the EDAW Brisbane office and provided the follow-up support for the case project. Furthermore, the new team members involved in construction supervision started to include staff with Asian backgrounds. These shifts suggest the progressive localisation of the EDAW team, a situation that peaked nearly eighteen months after their initial participation in the case project. A more locally-oriented material delivery thus started in the landscape design development stage. It was in this stage that the visual vernacular of the area were identified and finally materialised. This result contrasts with the international conceptual delivery, in which the proposals in response to local tradition and history were not successfully continued in the eventual design and design development phases.

This research suggests that, along with the progress from international conceptual to local material delivery, the relationship between the international practitioners and the other local stakeholder groups improved. As the case study showed, the intensive construction period paralleled the international practitioners' development of the landscape design. During this period, as discussed in Chapter Nine, substantial changes were made to the landscape design development package, including the entire shape of the proposed island. However, these significant design changes suggest a more direct engagement between the EDAW team and the local design team, which provided technical support and mediated with the local construction team. The case study showed that closer interaction was evident in the identification of the location for Li Dyke and the design of a traditional bridge, which were not included in the design development package provided by the EDAW team.

The locally-oriented material delivery was also evident in the reflections of the construction supervisor from the EDAW team. He (Brookbank 2007: personal communication) considered his supervision work less a formal construction management and more a process of influencing the local government clients to manage their own construction (Fig. 9.12). In this locally-oriented material delivery, a more direct engagement with local communities occurred. This construction supervisor (2007: personal communication) claimed:

> The best information was simply through on-going communication with the client and stakeholders... The greatest strategy one can use to understand the everyday life is to become a friend and understand the situation from heart... [As] you often find out more about local people at a banquet than you do from surveys.

This reflection suggests that the stage of material delivery tends to be characterised by less formal arrangements and liaisons. The less formal interaction also offered opportunities for the practitioners to gain a deeper understanding of the sense of place. Hybridisation naturally occurred as a direct result. It was evident in the emerging creativity in the later phase of international design practice, such as using new architectural forms and modern materials to interpret the visual vernacular, and the use

Fig. 9.12 Localising international practice during the design development and construction stages. *Source Urban Science Research of Wuxi: Li Lake Special* (Sun et al. 2007: 92)—the Australian landscape architect working on the construction site with the director of WMCPC (left) and the local design and construction teams (right)

of the artificial sandy beach to accommodate popular local activities. Other examples include adaptations to local materials and to manual operations, and the use of diverse, locally available plant species but in a new way of composition (Fig. 9.13, left). In a trip the author made for the case study, a local park user indicated to the author that the variety of plants was a pleasant new feature in the waterfront. These

Fig. 9.13 Plant design as shown in the constructed case site. *Source* Author, photographed in April and November 2007

hybridised elements coexist with some landscape and architectural forms popular in Australia, and with the paved plazas, flower boards and edged lawn areas that can be seen everywhere (Fig. 9.13, right). Such coexistence suggests that the combination of international conceptual and local material delivery constructed a landscape that is not simply characterised by hybridisation or homogenisation. Rather, it seemed to have resulted in an eclectic array of both in one physical space.

According to Daniels (2005), collective and long-term local interventions on an everyday basis are critical to the process of cultural hybridisation. Mitchell (2005: 188) also explained that hybridity is the outcome of this process and that a primary feature of hybridity is "integration and diffusion, of a thing that is derived from heterogeneous sources and composed of incongruous elements". The above argument suggests that the process of cultural hybridisation is a slow process, as it is through lived-in and long-term interaction. This parallels the slowness of gaining knowledge through personal engagement—a tacit form of knowing and exchanging knowledge between different contexts. It could be argued that the localisation of practice is an agent of cultural hybridisation, activated by skilled professionals that take part in the process on a long-term basis.

The case study showed there were integrations and diffusions engendered by the engagement between landscape architects and practitioners from Australia and worldwide, and between international practitioners and officials from local firms and institutions. The collectiveness and long-term local intervention were embodied in the materialisation of concepts, finally representing them in and as the constructed waterfronts parklands. This hybridity requires responsive and reflective practitioners of all sides. The case study also suggests that hybridisation may be more likely to occur in the context of design development and construction, but less likely in the conceptualisation of abstract space. Therefore, in order to allow hybridisation and to reduce homogenisation, strategies to acquaint international practitioners with local people are need to be developed for the international conceptual delivery, and prior to the establishment of branch offices in countries where services are to be provided.

9.3 Difficulties in Gaining Implicit Knowledge of Place

This research argued for the primary goal of landscape architecture to sustain the characteristics of place, the approaches to achieving this goal, and the multiple dimensions of gaining local understanding in international design practice. The previous discussions of the approaches to gaining knowledge of the Li Lake area and categorises of the identified characteristics suggest a general lack of understanding of meanings, ways of life, experiences and the collective memory that belong to the peasant community that lived in the Li Lake area. As Chapter Two argued, these are hidden aspects of place that convey community-based values and belong to implicit knowledge covering a particular place. This research identified two problems influencing international practitioners' acquirement of such understanding. One is reflected in ways to convey implicit knowledge of place to international practitioners. The other is reflected in the accessibility of international practitioners to the local communities, especially the peasants.

9.3.1 Ways to Convey Implicit Knowledge

This case study showed that current methods to inform international practitioners of the valuable but hidden aspects of place are problematic. First, the request for the practitioners to identify and incorporate meanings, cultural values, and habits of the local community was not explained clearly and was conveyed to the international practitioners in a relatively informal format. The case study found that the project brief showed the clear local demand for considering local ecological recovery and a stronger tourism economy. These formal and explicit requests contrast with the clients' messages concerning local history and tradition, which were communicated more verbally in the planning phase.

Second, the explicit request of integrating local history and culture in planning and design was after the international design teams' production of formal planning reports. As analysed in Chapter Eight, WMCPC saw themselves starting the task to integrate local culture and tradition 'early in the initiation of the project' (Sun et al. 2007: 18). This early initiation was marked by an invitation to local cultural groups to offer suggestions for the planning schemes. Guided by such an aim, a Cultural Symposium was held annually between the WMCPC and local cultural circles. The government saw this gesture as a reform in local urban development because previously such work had been conducted after the physical construction was complete (Sun 2007: 18). Although this reform suggests that the planning and design practice as culturally sensitive, knowledge of the non-material component of local history was not timely and sufficiently conveyed to international practitioners for planning purposes. In contrast, the case study showed ways to use local culture were oriented by economic purposes. As Chapter Two argued, the integration of cultural strategy (using history) with economic strategy (using technology) may lead to the image building of places. This, however, is evidenced in the constructed Li Lake waterfront.

Furthermore, knowledge of the valuable but hidden aspects of place was not sufficiently available. This lack was evident in the continuous emphasis on the heroic part of ancient local history, which was by both local stakeholders and international practitioners. The case study showed that in their early planning phase, the international practitioners focused on legends of ancient heroes and monumental evidences of local history, such as the ancient Grand Canal and urban structure. The case study showed that it was a result from the study of material provided by clients and experts and a less direct engagement with other local stakeholder groups. However, after the planning phase, and after the Cultural Symposium and the public participation, references to local history remained similar in the landscape design phase. It could be argued that past events, forces driving the evolution of the area, ways of life and the collective local memory, which theorists (Relph 1976; Tuan 1977; Meinig 1979; Hough 1990; Corner 1999; Girot 1999) see as valuable hidden aspects, remained unidentified and could not be conveyed to international practitioners as a type of formalised knowledge. It could even be argued that the readily documented heroic local past was the only source of formalised knowledge of local history that was conveyed to international practitioners.

Theory (Hough 1990; Paterson and Pettus 1994) suggests that the characteristics of place are shaped by the interwoven natural and social forces that drive a place's evolution. The identification of these forces requires a study of the successive layers of history and the discernment of significant but imperceptible site elements (Girot 1999). Theory (Seddon 1972; Hough 1990) also suggests that communities' collective interaction with the natural environment is a social force that promotes the evolution of a place. Furthermore, theory (Hebbert and Sonne 2006) suggests that knowing collective local memory is a sound historiographic strategy for contemporary planners and designers to provide environmental stewardship for the urbanising landscape. These theories show that the significance of the formless communal values and memories and forces is in ways of life and changes to place, more than in monumental entities and outstanding figures.

The research argues that it is less likely that international practitioners will identify and codify the abovementioned aspects of place by themselves. The formalisation of local knowledge by international practitioners themselves takes a long time, which conflicts with developers' tight schedules and the demands of international design practice as a private business. The main way that international practitioners know the history of place is through reading publications and documents, which is a process of gaining explicit knowledge of place (Lorch and Cole 2003), as explicit knowledge is 'codified' knowledge that is transmittable in formal, systematic language (Nonaka and Takeuchi 1995). It is thus very important for local researchers to identify the valuable but hidden aspects beforehand and to codify their research findings into explicit knowledge and convey this knowledge to international practitioners at the earliest stage of their involvement.

The research further suggests that international practitioners should directly engage with local researchers to gain this knowledge. As discussed in Chapter Two, a limiting factor to international design practice is the transfer of value systems from one culture to another (Lorch and Cole 2003: viii). This means that gaining knowledge through studying the readily-available materials may not be sufficient for international practitioners to understand how these hidden aspects of place are valued. King (2007) points out that in trans-national practice, there is an impetus to search for local history and tradition. In addition to search into publications and archival records, the research shows that the fulfilment of this ambition requires direct interaction with local researchers and effective communications.

Theory (Bull et al. 2007: 212–219) suggests that communicative methods, such as symposia, workshops and studios, are effective for exchanging knowledge, generating ideas and encouraging reflection on lived experience. The case study identified that workshops were adopted in the international design practice. However, these workshops were between international practitioners and the government developers, between government developers and local cultural circles, and between professionals of the culture of the international practitioners' own. The international practitioners and local researchers mainly communicated through meetings led by the government developers.

Ways for international practitioners to gain implicit knowledge of place, therefore, could be improved in three steps. First, local researchers would study the valuable but hidden aspects of place, initially identify this knowledge rather than leaving it to international practitioners' self-study and repeated visits. Second, such knowledge should then be expressed in an explicit format and effectively transmitted to international practitioners prior to their conceptualisation of the site and place. And third, international practitioners should directly engage with local researchers for a more effective understanding of different valuing systems. Figure 9.14 shows the model of gaining implicit knowledge of place that is adapted from theory. In this model, international practitioners conducting research of sites or places, codifying research findings into knowledge, adapting their valuing system into the local context. The process of gaining implicit knowledge to inform planning and design thus is over extended time.

Figure 9.15 shows that implicit knowledge is conveyed to international practitioners through a more active role played by local researchers and a closer interaction between the two cultural groups. The process of planning and design being informed by implicit knowledge of place is thus over limited time.

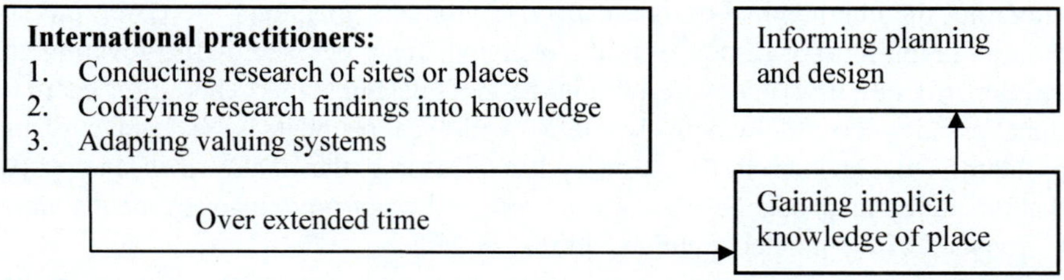

Fig. 9.14 Gaining implicit knowledge over extended time. *Source* Adapted from theory (Hough 1990; Paterson and Pettus 1994; Nonaka and Takeuchi 1995; Girot 1999)

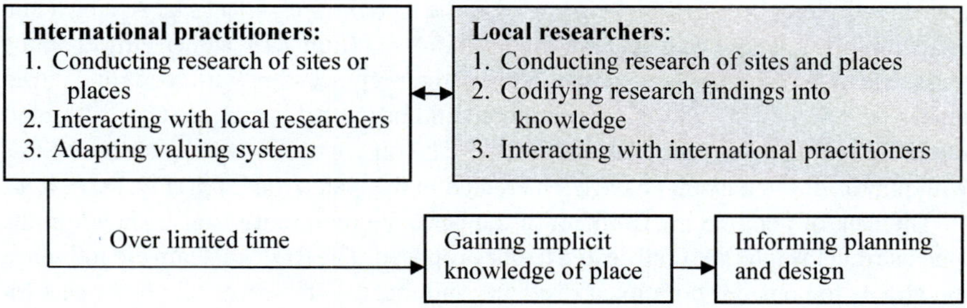

Fig. 9.15 Gaining implicit knowledge over limited time. *Source* Developed from the case study

9.3.2 Accessibility of Local Communities

This research found that there is insufficient access to local communities in the context of international design practice. This lack of access reduces the possibility of understanding communal values and the recent local life on the land that has been proposed for development. Theory (Seddon 1972; Relph 1976; Hough 1990; Seddon 1997) suggests that involving local communities is important for maintaining and retaining a sense of place. Hough (1990) also indicated that, in the development of places, involving the local community encourages emotional investment in places and can be a stepping stone to the emotional investment of tourists. He argued that this approach can associate the tourism economy with environmental ethics and reduce the destructive effect of tourism development. These goals, however, require a range of methods, such as conducting surveys, gathering documentary and oral histories, and public participation (Hebbert and Sonne 2006). Herbert and Sonne (2006) argued for the full deployment of these methods as taking on the role of an environmental steward and avoiding the risk of a one-sided sense of history. As the case study showed, the project adopted the method of public participation. However, the power of this method was reduced by three factors: less direct engagement, late public participation, and the lack of genuine involvement of the peasant community who were the original residents and interacted with the land and water on an everyday basis.

The first and second problems were evident in the involvement of local residents only after the international design teams had produced planning. Moreover, public opinion in the form of responses to the planning was processed by the government developer before it was conveyed to international design teams. Therefore, international practitioners did not actively interact with local residents by themselves. This problem is also evident in the claim by the director of the EDAW Brisbane office that the public feedback the EDAW team received was from the government's view (Harvison 2010: personal communication).

The lack of genuine involvement of the peasant community is shown in the constructed waterfront. As the case study showed, the government gave priority to the peasants who would lose their lands, houses and jobs in the development. This priority was expressed as guaranteeing new jobs for them and relocating them to new apartments along the proposed main road in the Li Lake area—Li Lake Avenue (Sun et al. 2007: 18). These apartments were built in 2007 (Fig. 9.16, right). However, the clearance of rice paddies and peasant houses, their replacement with roads and paved plazas, and the newly constructed sheltered and paved fishing areas (Fig. 9.10), left few traces of the lives that were once there. The only trace is the illustrative boards with photos of the past that have been erected in the waterfront area (Fig. 9.16, left).

The lack of genuine involvement of the peasant community is also shown in the land use that the international design teams proposed. The RDUDP team did influence the clients to consider both local residents and tourists. However, the proposals for local activities were applicable to many Chinese cities and towns. Proposals such as bicycle paths and children's playgrounds were not distinctive, nor did they respond

Fig. 9.16 The local peasant community—the old and new residences. *Sources* Left—author, photographed at the case site in November 2007, of the illustrative boards showing old peasantry houses originally at the site. Right—*Wuxi stepping toward an ecological water-mountain waterfront city: chronicles of the building of the new city of Li Lake* (Sun and Shao 2007: 42), showing newly constructed apartments for the peasant community

to the kinds of lifestyles exhibited by the local peasant population. Likewise, for the EDAW team, the clients' message on tourists overwhelmed that of local residents. The ideas proposed for a place oriented to tourism were based on Western everyday life, such as paved waterfronts, tidy fish markets, and water sports popular in the West. Presenting the area as culturally active living spaces reflected little of the authentic everyday life of the fishermen and peasants. An awkward relationship emerged over the struggle of international practitioners to make references to generic types of activities for Chinese people and to a more readily communicated heroic local past.

The case study showed the environmental expertise of Australian landscape architects, consistent with the claim of Armstrong (1999: 34)—an Australian educator in landscape architecture, that Australian landscape architectural practice has developed new ways to deal with issues such as pollution and water quality. However, the engagement with peasant community—the original rural residents of the land, was not evident in the case study, inconsistent with Armstrong's claim that the Australian landscape architectural practice today increasingly focuses on community interests.

The above inconsistency suggests that in the international context of design practice, foreign practitioners are apparently less likely to engage with local communities directly, especially with peasant communities and the old generation, as they remain less educated with English language. The reduced power of public participation in the context of international design practice suggests the need of new approaches. The research proposes two methods to improve the use of the strategy for public participation and the acquisition of knowledge regarding local communities. The first method uses local landscape architects as consensus builders to establish a common ground between the various local stakeholders, rather than as local technical supporters in a late stage of international design practice. In the second method, international practitioners introduce best-practice methods, including those for involving local communities, from their own countries to local researchers and practitioners, rather than cutting and pasting good outcomes alone. These approaches require an earlier collaboration between the landscape architects working locally and internationally, between international practitioners and local researchers, and direct ways of engagement, such as workshops, symposium, laboratories for testing new materials or technology for sustaining local artisanship, and conducting fieldwork together.

The following figures demonstrate the difference between the approach identified in the case study (Fig. 9.17) and that recommended by the research (Fig. 9.18). Figure 9.17 shows that opportunities for cross-cultural interaction were dominated by formal engagement such as meetings and searching into documents. The local researchers, as a part of local experts, did not have a strong voice in articulating community-based values of the Li Lake area in early stage. The local landscape architects remained passive providers of technical support in the construction stage. Finally, the planning and design were altered passively according to the emerging needs of the government and limitations of local supply of landscape materials, construction skills and tight schedules. In this process, values based upon the peasant community were overlooked.

Figure 9.18 shows that in addition to formal exchange of knowledge through searching into readily available publications and documents and organised meetings, diversified forms of engagement and interaction, such as tours and fieldwork aimed at immersion rather than sight-seeing, symposia, and workshops, provide opportunities for collaboration amongst international practitioners, local researchers and local landscape architects. The middle space of cross-cultural interaction thus becomes a

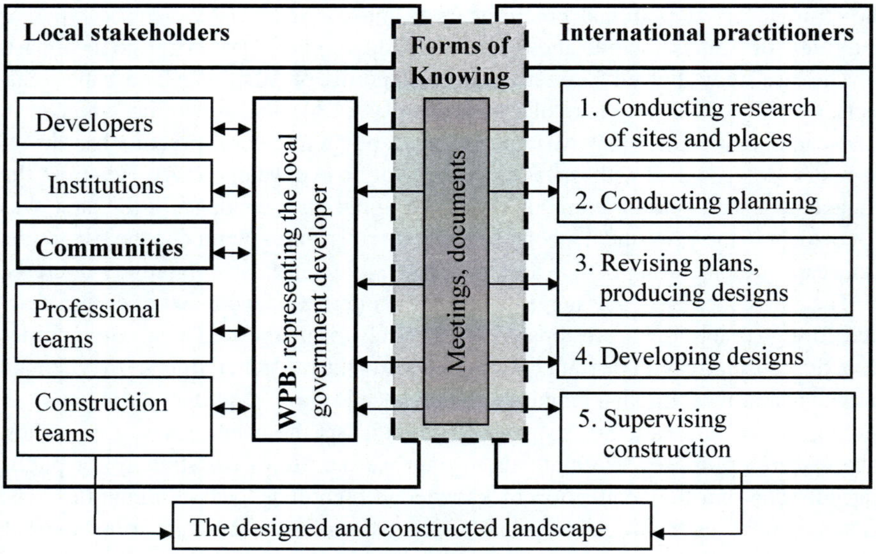

Fig. 9.17 Knowing local community through formal engagement with government. *Source* The case study

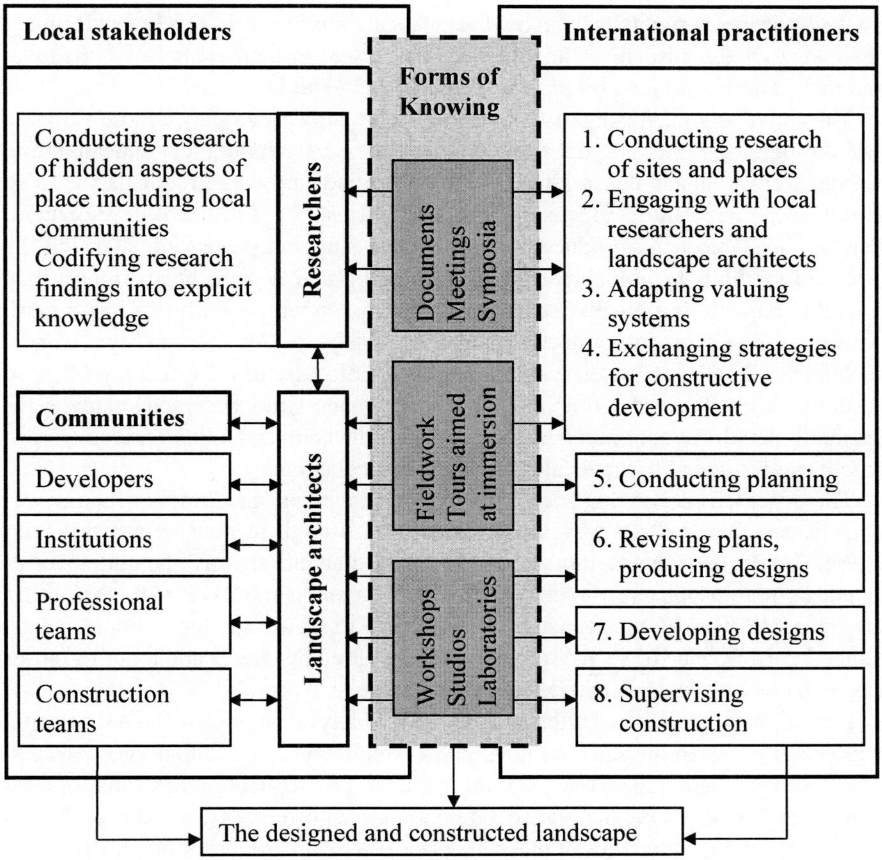

Fig. 9.18 Knowing local community through diverse forms of engagement with local researchers and landscape architects. *Source* Developed from theory and the case study

site of localising strategies and policies, planning and design ideas, and identifying practical methods to ensure good practice.

In this process, local researchers conduct research of hidden aspects of place, including the community-based values, codify research findings into knowledge, and provide this knowledge to international practitioners in the earliest stage of their participation. In addition to conveying such knowledge through documents, international practitioners engage with local researchers directly through workshops or symposia to gain further understanding and to adapt their valuing systems to the local context. International practitioners visit and investigate the site, and conduct fieldwork with local landscape architects together. Local landscape architects use their familiarity with the area to involve local communities directly and to build a common ground amongst diverse local stakeholders. In this second model, the collaboration between landscape architects working locally and internationally starts in the earliest stage of the international participation. Therefore, planning and design can be informed in timely fashion by diverse local opinions. As a result, the designed and constructed landscape is the outcome of a closer engagement between the local and international, bridged by local researchers and local landscape architects.

The above approach suggests that landscape architects can play a more valuable role through their international work. In the process of gaining local understanding through a close engagement with local researchers and landscape architects, they also introduce policies and strategies for good practice and constructive development. For instance, they can introduce policies of assessing and preserving sites and the practical methods to implement them in planning and design. Furthermore, they actively adapt these policies locally through work internationally. In this regard, Australian strategies for protecting and assessing places of cultural significance, such as the BURRA Charter, would result in a more careful identification of sites' aesthetic, scientific, social, cultural and historic values. Other accepted strategies for engaging with local communities should be introduced through international design practice and adapted to the local social and political context.

This research asserts that such experience is necessary in Chinese urban development and that the best way to introduce it is through responsive practitioners' international work. This method is more effective than introducing planning theories and design philosophies from developed countries and is more scientific than using the image of world-famous places as benchmarks. However, international practitioners' performance of such a role must be supported by clients and requires closer engagement with local researchers and practitioners.

Although the BURRA Charter is being more widely considered in the Asia-Pacific region and has been introduced into China's Architecture educational programmes, it has not been generalised to education in landscape architecture in China. In relation to the CHSLA's debate and its comparisons with the West for approaches to sustaining the characteristics of place in development and urbanisation, this research further suggests comparisons of practice. This means that in addition to discussing environmental expertise and design philosophies, the topic of formalising values of hidden aspects of places, and interpreting them in landscape architectural planning and design, should be included in the professional debate and discussed across cultures.

References

AILA. (2005). Code of professional conduct. Retrieved October 3, 2010, from http://www.aila.org.au/codeofconduct/.

Armstrong, H. (1994). Inventing Landscape: New Collabrative Design. The culture of landscape architecture. In H. Edquist & V. Bird (Ed.) (pp. 117–134). Melbourne, Vic., Edge Publishing in association with the Department of Planning Policy and Landscape RMIT.

Armstrong, H. (1999). Landscape design in Australia. *Suzhou Institute of Urban Construction and Environmental Protection, 12*(3), 30–34.

Arvidson, A. R., & Martin, F. E. (2007). Changing places: where history and nature collide. *Landscape Architecture, 97*(11), 48–57.

Australia Parliament Senate Foreign Affairs Defence and Trade References Committee & Forshaw, M. (1996). *Australia China relations*. Canberra: Parliament of the Commonwealth of Australia.

Brookbank, P. (2007). Interview with Yun Zhang on 06 November at EDAW Shanghai office.

Bull, C., Boontharm, D., et al. (Eds.). (2007a). *Cross-cultural urban design: global or local practice?*. Routledge: London.

Bull, C. J., Radovic, D., et al. (2007b). Urban design for a cross-cultural future. In C. Bull, D. Boontharm, & C. Parin (Eds.), *Cross-cultural urban design: Global or local practice?* (pp. 208–233). London: Routledge.

Corner, J. (Ed.). (1999). *Recovering landscape: essays in contemporary landscape architecture*. Princeton Architectural Press: New York.

Corner, J. (2006). Terra Flux. The landscape urbanism reader, 21–33.

Daniels, P. W. (Ed.). (2005). An introduction to human geography: issues for the 21st century. Prentice Hall: Harlow.

EDAW (2005). 2005 Waterfront Awards: Wuxi Li Lake Waterfront Parklands.

Editorial Board of *China Wuxi* Journal. (2002). "Brief introduction of detailed planning of Li Lake New City." Retrieved May 7, 2008, from http://www.wuxi.gov.cn/intowx/survey/admin/20069/20069527768.shtml.

Fu, Y. (2007). "EDAW in Asian Market: 1997–2007." Retrieved July 16, 2008, from http://sh.focus.cn/news/2007-06-29/330133.html.

Frampton, K. (1983). *Seven characteristics of critical regionalsim*.

Girot, C. (1999). Four Trace Concepts in Landscape Architecture. *Recovering landscape: essays in contemporary landscape architecture*. New York, Princeton Architectural Press: pp. 59–67.

Guo, H., & Li, Y. (2007). Habitant, picturesque scene and artistic conception: landscape design of northern shore of Jincheng Bay. *Hua Zhong Architecture, 25,* 81–91.

Harvison, G. (2007). Interview with Yun Zhang on 3 October at EDAW Brisbane' office.

Harvison, G. (2010). Communication through email, sent to University of Melbourne, on 25 May 2010.

Hebbert, M., & Sonne, W. (2006). History builds the town: on the uses of history in twentieth-century city planning. In J. Monclus & M. Guardia (Eds.), *Culture, urbanism and planning* (pp. 3–19). Aldershot, England; Burlington, VT: Ashgate.

Hough, M. (1990). *Out of place: restoring identity to the regional landscape*. New Haven: Yale University Press.

King, R. (2007). Confronting epistemes. In C. Bull, D. Boontharm & C. Parin (Ed.), *Cross-cultural urban design: Global or local practice?* (xvii–xxvi). London: Routledge.

Lorch, R., & Cole, R. (Eds.). (2003). *Buildings, culture and environment: informing local and global practices*. Oxford, UK: Malden, MA, Blackwell Pub.

Lin, Q., & Wang, X.-R. (2005). Regional features and landscape form. *Chinese Landscape Architecture, 21*(114), 16–24.

Miao, P. (1989, 1991). Essences of tradition—thirteen characteristics of Chinese architecture. *Architects 36, 40*(12,3), 56–67, 61–80.

Nian, y.-g. (2007). Technical renovation for ecological recovering of Li Lake. Urban Science Research of Wuxi: Li Lake Special z. Sun, J. Liu, x. He and m. Xu. Wuxi, Editorial board of Wuxi urban science research, 89, 150.

Nonaka, I., & Takeuchi, H. (1995). *The knowledge-creating company*. New York: Oxford University Press.

Norberg-Schulz, C. (1980). *Genius loci: Towards a phenomenology of architecture*. London: Academy Editions.

Meinig, D. W. (Ed.). (1979). *The Interpretation of ordinary landscapes: Geographical essays*. New York: Oxford University Press.

Paterson, G., & Pettus, A. T. (1994). The visible landscape. The culture of landscape architecture. In H. Edquist & V. Bird (Ed.) (pp. 139–154). Melbourne, Vic., Edge Publishing in association with the Department of Planning Policy and Landscape RMIT.

Relph, E. C. (1976). *Place and placelessness*. London: Pion.

Seddon, G. (1972). *Sense of place: a response to an environment, the Swan coastal plain Western Australia.* Nedlands: University of Western Australia Press.

Seddon, G. (1997). *Landprints: reflections on place and landscape.* Cambridge; Melbourne: Cambridge University Press.

Sun. (2007). Interview with Yun Zhang on 27 December at the office of Wuxi's Municipal Committee of the Planning and Construction of Lihu City.

Sun, Z., & Shao, G. (Eds.) (2007). Wuxi stepping toward an ecological water-mountain waterfront city: chronicles of the building of the new city of Li Lake Wuxi, Wuxi's Municipal Committee of the Planning and Construction of Lihu City.

Sun, Z., Liu, J. et al. (Eds.). (2007). Urban Science Research of Wuxi: Li Lake Special Wuxi, Editorial board of Wuxi urban science research.

Tuan, Y.-F. (1977). *Space and place: The perspective of experience.* Minneapolis: University of Minnesota Press.

Wall, A. (1999). Programming the urban surface. In J. Corner (Ed.), *Recovering landscape: Essays in contemporary landscape architecture* (pp. 233–250). New York: Princeton Architectural Press.

Wuxi Government. (2004, Aug 30). Approaches and experiences of making a Model City of Wuxi. Retrieved June 20, 2008, from http://big5.mep.gov.cn/gate/big5/www.mep.gov.cn/cont/mhcity/mfcsxx/zyzfjy/200408/t20040830_60884.htm.

Wuxi Municipal Committee of the Planning and Construction of the Li Lake area, Ed. (2007). Unveiling the beauty of Li Lake: Wuxi well on its way towards ecological well being Phoenix Press Group and Phoenix Publishing House.

Wulff, R. (2007). Interview with Yun Zhang on 25 September, at Tract consultants' Melbourne office.

Wulff, R. (2010). Communication through telephone, called to University of Melbourne, on 12 May 2010.

Yang, z.-y. (2007). Conversation with Yun Zhang on 27 December at the office of Wuxi's Municipal Committee of the Planning and Construction of Lihu City.

Young, G. (2006). Speak, Culture!—Culture in Planning's Past, Present and Future. *Culture, urbanism and planning* (pp. 43–59).

Chapter 10
Conclusions

The previous chapter discussed the position of landscape architectural practice in the international context and Australian landscape architects' approaches to place and practice in China. The characteristics of international landscape architectural practice and landscape architects' approaches to engaging with the characteristics of place are thus defined in a number of ways. The previous chapter also discussed the difficulties in gaining implicit knowledge covering a particular place. Such discussions disclose the complexity of culturally responsive landscape architectural practice in the context of providing planning and design services internationally. Such complexity also suggests important directions for future research.

10.1 The Characteristics of International Landscape Architectural Practice

The first research question of this research asked what the characteristics are of Australian landscape architectural practice internationally and specifically, in China. The research concludes that the international practice of Australian landscape architecture has been a product of the process of urbanisation in Asian developing countries. Specifically, such international landscape architectural practice has been active in the development of China's urban fringe, essentially holistic in providing multidisciplinary services, and tended to increase more rapidly in the contemporary situation.

International landscape architectural practice has been valued on the ground of landscape architects' environmental expertise. Such expertise is manifest in building a balance between new urban use and the enhancement of the natural characteristics of original rural landscapes. The fact that landscape architecture is centred on planning and design of the land and its management has made the profession a

Y. Zhang, *Place Making in International Practice of Landscape Architecture*,
Advanced Topics in Science and Technology in China 64,
https://doi.org/10.1007/978-981-16-2442-1_10

potentially important player across the globe, especially in Asian countries where vast urbanisation is occurring and rural lands are transformed into built environments.

In the last decade, Australian landscape architects have expanded their working territories from Southeast Asia to East Asia and from the design of inner urban spaces and residential districts in China to the planning of land use in urbanising fringes. Such a shift has been promoted by China's reform and the consequent urbanisation, socio-cultural change and market conditions of planning and design. The recognition of international planning and design practice as landscape architectural practice in China has been promoted by such a form of environmentally sensitive practice. A culturally sensitive international practice in landscape architecture, however, requires the association between the environmental expertise of international practitioners and the cultural habits of the local community in which services are provided. The central task of landscape architectural practice internationally thus becomes two-fold. One task is to express meaningfully the scientific basis of local ecological recovery. The other is to cultivate places that are compatible with activities not only of new comers (tourists and international investors), but also the lasting activities of local communities.

The types of international landscape architectural practice have extended from land use planning, design, construction supervision, and even to landscape maintenance, and the education system of the local profession. The significant participation of landscape architects in changing land from rural to urban use has resulted in a broader recognition of landscape architectural practice in the international context. Landscape architects have provided leadership in the complexity of providing services and have played a key role within hybrid international design teams. The extension of working territory and holistic participation in local development suggests that in the future, there is to be longer and deeper impacts upon local developmental planning and design. Yet, conversely, it has also been found that many opportunities for localising international design practice are yet to be capitalised upon.

International landscape architectural practice has become a means to connect the local profession of landscape architecture with the international movement. The increase in landscape architectural practice internationally, over time, promoted the growth and the institutionalisation of Chinese landscape architects and enhanced the association between them and the International Federation of Landscape Architecture (IFLA). In the present phase of global urbanisation, international landscape architectural practice would tend to a wider collaboration among countries and across cultures. The geographical and cultural positions of Australia can promote Australian landscape architects to the frontier of this wider collaboration. Australian landscape architects can play a more active role in this collaboration and act as consensus builders in an international context in a number of ways: first, by building a stronger network amongst Australian landscape architects working internationally; second, by engaging in the establishment of a code of professional ethics with the local profession and clients.

10.2 The Approaches to Engaging with the Characteristics of Place

The second research question of the research asked how Australian landscape architects engage with the characteristics of place when they practiced in China. Four approaches typical in the context of international planning and design practice were defined. The first was identifying effective methods of operating a practice in other cultures through collective knowledge and previous experience. This resulted in hybrid forms of international design teams that were composed of different firms or offices. The second was key practitioners' intensive investigations of sites and periodic visits for generating planning and design concepts. This formed the international conceptualisation of place. The third was building partnership with clients, receiving information from them and using them as stepping stones to engage with other local stakeholders. This provides international practitioners priorities in delivery planning and design concepts to the ground. The fourth was localising practice through building branch offices, posting staff on long-term bases, and directly employing staff with Asian backgrounds. This contributed to the locally oriented materialisation of place.

In the early stages of the international design teams' involvement, the conceptualisation of planning and design required collective knowledge combined with key practitioners' intensive investigation of sites and the building of partnership with clients. Ultimately, such actions led to the acquirement of explicit knowledge of place, based upon the visual vernacular, the heroic local past and generic types of activities, and this knowledge was used as a reference to the characteristics of the Li Lake area in planning and design. In the international conceptualisation, the gaining of implicit knowledge of the area was hindered by the format and quality of local provision of such knowledge, by the limited time allowance for international practitioners to formalise such knowledge by themselves, and by the limited access to local communities. The gaining of implicit knowledge of the area only began at a late stage when landscape architectural practice became more localised. At this stage, the planning and design proposals were altered along with the identification of local limitations in materials and construction skills and the emerging decisions of the government.

10.3 Propositions for Quality International Planning and Design Practice

International practice in landscape architecture has been seen to become broader in its scope of providing services, deeper in its involvement in local development, and rapidly increasing. However, in the context of such findings, the opportunities for international practitioners to engage with sites and local stakeholders fail to satisfy the acquiring of implicit knowledge covering a particular place. However, there are a number of ways to correct this situation so that practitioners, clients, researchers and educators can act to support quality international practice.

First, this research proposes that international practitioners build partnerships with local academia and local practitioners. In doing so, they would actively introduce conservation policies and practical methods from their own experience and countries; and adapt their experience into local systems through working with local practitioners and researchers together. Specifically, local habits of using outdoor space, meanings associated with plant species, and local materials and artisanship should be identified and considered in the conceptual planning and design, and through such collaboration.

Second, Chinese clients who procure design services internationally need to be more aware of the importance of communicating cultural difference. As a key local supporter of international practitioners, quality international design practice relies a great deal on clients, especially in practices for large land change in which clients are often positioned as the top local stakeholder, the government. Clients should understand that cultural significance of sites and landscape is often in an implicit form. Clients should assess the availability of knowledge of the valuable but hidden aspects of place, prior to proposing change, and should invite international design teams to participate in sharing such knowledge. Clients should also understand that the best way to use the experience and knowledge of international practitioners is to adapt their experience and strategy into local systems. Therefore, clients should actively offer opportunities for international practitioners to conduct such adaptation. Given the often limited time for development, clients should offer international practitioners earlier and direct engagement with local researchers and local practitioners, prior to conceptual planning and design.

A third proposition is aimed at researchers and educators. The research identified the significance of implicit knowledge of place that has been argued for by researchers and design scholars, but remains less accessible than explicit knowledge in the context of the everyday working of international design practice. The research proposes that researchers support quality international practice in three ways. The first is by developing research into the international practice of planning and design, which is necessary in our globalised world. The second is to act as interlocutors between local developers and international practitioners to conduct cross-cultural studies of heritage conservation and to develop ways of articulating local context. The third is to develop education programs for landscape architecture. This could be the education of knowledge essential for conducting work internationally and the evolution of professional knowledge on global practice. Education programs should also emphasise cultural nuances embedded in human interactions with the natural environment, and ways of gaining and using implicit knowledge in practice. Although the trends indicate that developed Western societies have more actively engaged in the international design market, clients and local practitioners of developing countries are also crucial to quality international design practice. Therefore, such education programs should be developed in both the demand and supply sides of international design practice: in both Australia and China.

This research produced a new model (Fig. 9.18) for expanding the space for knowledge exchange through international planning and design practice including landscape architecture. This model emphasises personal and informal engagement between local stakeholders and international practitioners, highlighting a tacit form of knowing for a cross-cultural realm. The model also encourages international practitioners' intensive and holistic collaboration with local researchers and local practitioners, aiming at sharing strategies for environmental challenges that present-day places are facing.

10.4 Directions for Future Research

The research is a beginning in the investigation of the practical work of landscape architectural practice internationally, through its focus on Australian landscape architectural practice internationally and in China. This research defined the characteristics of international practice in landscape architecture and its approaches to engaging with local characteristics of place. The findings of the research suggest directions for future studies in relation to international design practice and to landscape architectural theory.

Future studies in relation to international design practice can develop in three ways. First, as this research focused on the work and perspectives of international practitioners, the identified problems were seen mainly on the local demand side. Future research can develop a method to conduct in-depth investigations into the relevant work and perspectives of clients who form a vital component of cross-cultural engagement. Second, this research referred to Australian landscape architectural practices internationally and in China. The specificity of Australia as a supplier of international practice can be further explored in relation to the history of the profession, and to Australia's multicultural context that supports its international practice in Asia. These two areas would further expose the specific characteristics of Australian landscape architectural practice in Asia. Third, this research selected the domain of Western design practice in non-Western countries. Future study can adapt the method of this research to study the other two notable areas of international design practice: first, between Western societies, such as US practice in Australia; and second, between non-Western societies, such as Japanese practice in China. These two areas of future study would contribute to a more holistic understanding of international practice in landscape architecture. Finally, the designed landscape by international practitioners needs ongoing monitoring, particularly with regard to how various socio-economic groups, such as China's peasant population, who are most affected by land change and urbanisation, adapt to change.

In relation to landscape architectural theories, further research could explore the interrelationship between environmental ethics and landscape preference. The findings of such research might support the construction of meaning onto scientific strategies for ecological recovery. They can also support planning as both environmentally and culturally sensitive, especially useful for the cross-cultural types of planning services now requested of landscape architecture. Such research can perhaps provide a stepping-stone to allow crossing the gap between the scientific thinking associated with planning and the material thinking and art creation of landscape design.

Appendix A

Survey Questionnaire

Q1. The size of your company

Please indicate the number of full-time employees in your company (please tick appropriate box)

June 1995	☐ 0-19	☐ 20-49	☐ 50+
June 2000	☐ 0-19	☐ 20-49	☐ 50+
June 2005	☐ 0-19	☐ 20-49	☐ 50+

Q2. International practices from 1995 onwards

Based on the approximate service fee of all projects, please rank the contribution of international projects on a scale of 1 to 5 of total fee received, where 1 is 'less than 10%' and 5 is 'more than 50%' (please tick appropriate box)

	1 <10%	**2**	**3**	**4**	**5** > 50%
To 1995	☐	☐	☐	☐	☐
1995–2000	☐	☐	☐	☐	☐
2000–2005	☐	☐	☐	☐	☐
2005–2007	☐	☐	☐	☐	☐

Y. Zhang, *Place Making in International Practice of Landscape Architecture*, Advanced Topics in Science and Technology in China 64,
https://doi.org/10.1007/978-981-16-2442-1

Q3. Landscape practices in China from 1995

Please rank the contribution of Chinese projects to the total fee for all international projects on a scale of 1 to 5, where 1 is 'less than 10%' and 5 is 'more than 50%' (please tick appropriate box)

	1 <10%	2	3	4	5 > 50%
To 1995	☐	☐	☐	☐	☐
1995–2000	☐	☐	☐	☐	☐
2000–2005	☐	☐	☐	☐	☐
2005–007	☐	☐	☐	☐	☐

Q4. Type of landscape projects in China

Please indicate the type of projects in China you have worked on between 1995–2000 and 2000–2005, and please rank each type of projects from 1 to 7, where 1 is 'the least' and 7 is 'the most' projects. (Please tick appropriate box, and write appropriate number between 1 to 7 in the last column)

** G & P: joint investment of government and private sector*

1995–2000 **Project_ site**	**Client** Government	 Private	 * G & P	**Scale** 1-7
Residential district/estates	☐	☐	☐	☐
Urban precinct/district	☐	☐	☐	☐
Inner urban public space	☐	☐	☐	☐
Urban fringe public space	☐	☐	☐	☐
Urban fringe industrial park	☐	☐	☐	☐
Rural scenic spot	☐	☐	☐	☐
Other, please specify:	☐	☐	☐	☐

2000–2005 Project_ site	Client Government	Private	* G & P	Scale 1-7
Residential district/estates	☐	☐	☐	☐
Urban precinct/district	☐	☐	☐	☐
Inner urban public space	☐	☐	☐	☐
Urban fringe public space	☐	☐	☐	☐
Urban fringe industrial park	☐	☐	☐	☐
Rural scenic spot	☐	☐	☐	☐
Other, please specify:	☐	☐	☐	☐

____

Q5. How did you obtain projects in China between 1995–2000 and 2000–2005? Please indicate on a scale of 1 to 5, where 1 is 'the least' and 5 is 'the most' important (please tick appropriate box and write appropriate number between 1 to 5 in the last column)

** I: international competition, C: direct commission from client*

1995–2000	* I	C	Other	Scale 1–5
Prior relationship with other consultants	☐	☐	________	☐
(Such as: ____________________________________)				
Joint projects with other consultants	☐	☐	________	☐
(Such as: ____________________________________)				

Recommendation ☐ ☐ ________ ☐
(Such as from previous clients, or: ________________________________)

Openly sourced information ☐ ☐ ________ ☐
(Such as project known form client's advertisement, your company known from website, or________________________________)

Other, please specify: ☐ ☐ ________ ☐
____

2000–2005	* I	C	Other	Scale 1–5
Prior relationship with other consultants	☐	☐	________	☐

(Such as: ________________________________)

Joint projects with other consultants ☐ ☐ ________ ☐
(Such as: ________________________________)

Recommendation ☐ ☐ ________ ☐
(Such as from previous clients, or: ________________________________)

Openly sourced information ☐ ☐ ________ ☐

(Such as project known form client's advertisement, your company known from website, or________________________________)

Other, please specify: ☐ ☐ ________ ☐
____

Q6. Please indicate appropriate local providers that support your landscape practice in China during different stages (Please tick appropriate box), **and indicate how you communicated with them.**

* *C: client; LP: local professional; LI: local interpreter; LB: local builder*

* *Communication: eg. Branch office; Periodical visit; E-communication…*

1995–2000 Activity	**Local provider** * C	LI	LP	LB	*** Communication**
Pre-design services	☐	☐	☐	☐	________
Schematic design	☐	☐	☐	☐	________
Design development	☐	☐	☐	☐	________
Construction documentation	☐	☐	☐	☐	________
Construction supervision	☐	☐	☐	☐	________
Contract administration	☐	☐	☐	☐	________
Landscape maintenance	☐	☐	☐	☐	________
Other activity	☐	☐	☐	☐	________

Please specify: ________________________________

2000–2005 Activity	**Local provider** * C	LI	LP	LB	*** Communication**
Pre-design services	☐	☐	☐	☐	________
Schematic design	☐	☐	☐	☐	________
Design development	☐	☐	☐	☐	________
Construction documentation	☐	☐	☐	☐	________
Construction supervision	☐	☐	☐	☐	________
Contract administration	☐	☐	☐	☐	________
Landscape maintenance	☐	☐	☐	☐	________
Other activity	☐	☐	☐	☐	________

Please specify: ________________________________

Q7. Do you actively seek to expand your landscape practice in China? (Please tick appropriate box)

Yes ☐ No ☐

Q8. Are there any challenging aspects of international practice in China? Could you describe your strategies in dealing with these challenges? (Optional)

____

____

____

____

____

____

____

____

Thank you again for your time.
HREC NUMBER: 0709877.1 August 29, 2007

Appendix B

Interview Questions

Open question 1:

Could you describe the most impressive aspects of your professional experience of (name of participant project) in China, which are different from those in Australia?

Follow-up questions:

a. Which were the aspects of project productions that most influenced the landscape design outcome in this project? How do they differ from the kinds of influences that might occur in Australia?
b. Were there any preferences of aesthetic outcome, public use, or landscape maintenance of clients or other consultants revealed in this practice in China?
c. Were there any operational differences in terms of the method of delivery, techniques used to represent the landscape design, or the process of construction?

Open question 2:

How do you consider the natural character around Lake Tai, the biggest fresh water lake in China, in your work of (name of participant project)?

Follow-up questions:

a. What kinds of method, technique or instrument did you use to collect information of the natural environment of the particular site you worked on?
b. How do they differ from your approaches in Australia?

Open question 3:

How do you consider the everyday life and the tradition of local people of Wu Xi, in your work of (name of participant project)?

Y. Zhang, *Place Making in International Practice of Landscape Architecture*,
Advanced Topics in Science and Technology in China 64,
https://doi.org/10.1007/978-981-16-2442-1

Follow-up questions:

a. What kinds of method, technique or instrument did you use to collect such information? How do they differ from your approaches in Australia?
b. Which aspects of local everyday life do you think are most concerned with your work? How do they differ from your consideration in Australia?
c. Which aspects of local tradition do you think are most concerned with your work? Do you think the local tradition of literati garden matters to your work?
d. What do you think of the remnants of previous landscape, and their cultural aspects, such as historical legend? How do you integrate them into your work?

Open question 4:

Could you describe images of designed landscapes in China you had in your mind before you got involved in landscape projects in China?

Follow-up questions:

a. How did you get these images, where from and through what kind of media?
b. Were there any divergences from perceptions about China that you may have had before starting work there?
c. What do you think of the differences and what aspects were valued by you?

HREC NUMBER: 0709877.1 August, 2007

Appendix C

Appendix C: the Li Lake area in 2001

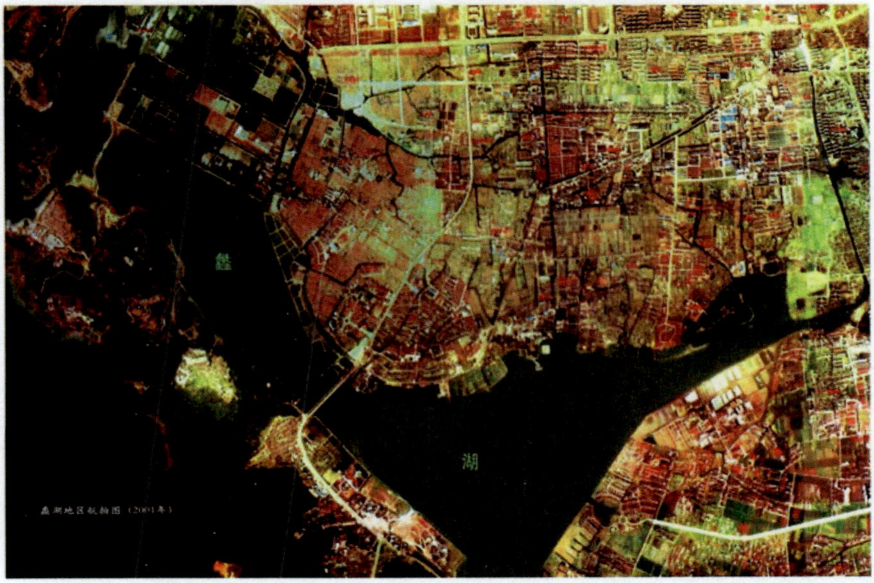

Y. Zhang, *Place Making in International Practice of Landscape Architecture*,
Advanced Topics in Science and Technology in China 64,
https://doi.org/10.1007/978-981-16-2442-1

Appendix D

Appendix D: the Li Lake area in 2007

Y. Zhang, *Place Making in International Practice of Landscape Architecture*, Advanced Topics in Science and Technology in China 64,
https://doi.org/10.1007/978-981-16-2442-1